Self-Organization in Continuous Adaptive Networks

RIVER PUBLISHERS SERIES IN INFORMATION SCIENCE AND TECHNOLOGY

Consulting Series Editor

KWANG-CHENG CHEN
National Taiwan University
Taiwan

Information science and technology enables 21st century into an Internet and multimedia era. Multimedia means the theory and application of filtering, coding, estimating, analyzing, detecting and recognizing, synthesizing, classifying, recording, and reproducing signals by digital and/or analog devices or techniques, while the scope of "signal" includes audio, video, speech, image, musical, multimedia, data/content, geophysical, sonar/radar, bio/medical, sensation, etc. Networking suggests transportation of such multimedia contents among nodes in communication and/or computer networks, to facilitate the ultimate Internet. Theory, technologies, protocols and standards, applications/ services, practice and implementation of wired/wireless networking are all within the scope of this series. We further extend the scope for 21st century life through the knowledge in robotics, machine learning, cognitive science, pattern recognition, quantum/biological/molecular computation and information processing, and applications to health and society advance.

- Communication/Computer Networking Technologies and Applications
- Queuing Theory, Optimization, Operation Research, Statistical Theory and Applications
- Multimedia/Speech/Video Processing, Theory and Applications of Signal Processing
- Computation and Information Processing, Machine Intelligence, Cognitive Science, and Decision

For a list of other books in this series, please visit www.riverpublishers.com

Self-Organization in Continuous Adaptive Networks

Anne-Ly Jabusch Do
Max Planck Institute for the Physics of Complex Systems, Germany

and

Thilo Gross
University of Bristol, UK

River Publishers

Aalborg

ISBN 978-87-92329-45-5 (hardback)

Published, sold and distributed by:
River Publishers
P.O. Box 1657
Algade 42
9000 Aalborg
Denmark

Tel.: +45369953197
www.riverpublishers.com

Contents

Introduction

One of the greatest challenges in physics, and perhaps in all science, is understanding *emergence*, the appearance of systemic properties and functions from the interaction of the microscopic parts of a system [1].

Many beautiful examples of emergence can be observed in nature [2]. Indeed, life itself can be regarded as an emergent phenomenon, as it is not an inherent property of the molecules that make up our bodies, but instead arises from their interactions. Likewise, our intellect is not derived from intelligent cells, but emerges from the interaction of many cells, which in isolation cannot be considered intelligent. The analogous notion of collective intelligence captures the observation that groups of animals often perform better than any individual in the group in isolation. Perhaps, the most amazing example of this collective intelligence is human culture. Clearly culture is not a property that is inherent in any single human, but emerges from the interactions between us.

Examples of emergence can also be found in artificial systems. However, there is a difference. While emergence in nature is mostly beneficial, supporting the function of the system, prominent examples in artificial systems include traffic jams, market crashes, and power cuts, which disrupt functionality. Where does this difference come from? Consider that present-day artificial systems are the product of a rational design process which prescribes the exact way in which functioning is achieved. Functioning then relies on the prescribed interaction between all parts of the system with narrow error margins. For example the millions of parts in industrial machinery require thoughtful design and exact assembly. Any emergent dynamics other than the desired function are in general not anticipated in the design process, and threaten to disrupt the delicate order on which the function is built.

Also natural systems require some degree of collective order to function. However, this order is often itself an emergent property that arises from a process of *self-organization*. The defining property of self-organization is the spontaneous generation of order from homogeneous or random initial conditions, based on local rules that do not require centralized leadership. Because self-organization does not rely on a predefined organizing center it can often continue to operate even when a significant fraction of parts of the system have failed. Because it is an ongoing process that is not strongly dependent on initial conditions it can often mitigate, repair, or overcome disturbances affecting the system. Self-organization can thus enable biological systems to survive serious perturbations that would derail a designed artificial systems.

An important goal in engineering is to harness the benefits of self-organization also for technical systems and possibly organizational structures. Present visions include for instance smart power-grids that can adapt to a rapidly fluctuating supply and demand generated by decentralized network of power-sources and sinks; decentralized traffic control that uses collective intelligence to optimize traffic flow; and self-organizing micro-processors that can remain functional even as a considerable fraction of their transistors have failed.

Designing a self-organizing system requires finding a set of local rules that give rise to desired system-level dynamics. In biological systems such sets of rules have been found through the process of evolution. One plausible approach is therefore to design technical systems by simulating such an evolutionary process [3]. However, evolutionary optimization is numerically demanding and is futile for the hugely complex technical system which would profit most from self-organization. This defines a strong demand for theoretical insights that can speed up optimization or replace it by a rational design process for self-organizing complex systems.

Progress in the rational design of self-organizing systems will most likely emerge from advances in complex systems theory leading to a heightened understanding of emergence. Presently, there is no unified approach that allows to trace system-level phenomena back to the properties of a system's constituents, nor a recipe for designing the constituents such that desired system-level behavior emerges. These missing links between the level of the part and the level of the system, limits our ability to both rationally design emergence in artificial systems and understand self-organization observed in nature. However, future progress will be able to build on recent advances in the three pillars of complex systems theory: statistical physics, nonlinear dynamics, and network science.

Statistical physics attacks complexity top-down, aiming to find agent-level explanations for observed system-level behavior. While most progress in statistical physics is still restricted to systems in or close to thermodynamic equilibrium, some recent advances in the analysis of non-equilibrium systems have been made [4, 5]. More importantly, statistical physics has developed tools for characterizing the macro-behavior of systems [6, 7] and provides a robust framework of concepts. Therefore terminology from physics has become essential for discussing emergence in a wide spectrum of disciplines.

Nonlinear dynamics provides a complementary approach, attacking complexity bottom-up by exploring the system level consequences of a given set of agent-level rules. Much attention in nonlinear dynamics has focused on low-dimensional systems [8], that cannot be considered complex according to the definition used in this work. However, in some areas key insights have been gained that extend to complex systems. One example is the Turing instability [9, 10] which describes a mechanism for the formation of patterns in systems of partial differential equations that can be understood analytically. This insight has significantly advanced the understanding of pattern formation in biology [11, 12] and the design of artificial systems that show self-organized pattern formation.

Network science has over the past decade played a central role in the investigation of complex systems. Describing a given complex system as a network means simplifying the complexity of the parts of the system, but retaining the complexity in the pattern of their interactions. Thereby networks provide a level of description that achieves a considerable simplification without oversimplifying the patterns of interactions that are essential for emergence. Major results concern for instance the giant component transition [13], the formation of scale-free topologies [14], and their implications for error and attack tolerance [15, 16].

It is conceivable that combining network science with concepts from statistical physics and methods from nonlinear dynamics can lead to a detailed analytical understanding of emergence that will prove essential for the future analysis and design complex self-organizing systems. One area of network science which has over the past decade already profited immensely from ideas from statistical physics and nonlinear dynamics is the study of adaptive networks [17, 18]. These networks combine a dynamical process taking place on the network with topological evolution of the network structure. The interplay of the two different dynamics leads to a rich variety of dynamical phenomena including highly-robust self-organized criticality [19], the spontaneous emergence of different classes of nodes from a homogeneous population [20], and new types of dynamics and transitions [21–23]. After simulation studies have opened up a plethora of interesting phenomena [20, 22–41], it is today's challenge to develop analytical approaches for addressing the underlying principles.

In the past the analysis of adaptive networks has profited from classical approaches of network science, which played for instance a key role in understanding percolation transitions in adaptive network [16]. Statistical physics has provided important underlying concepts, for instance for understanding self-organized criticality in adaptive networks [19]. And, finally tools from nonlinear dynamics are used frequently to explore the dynamics and transitions of networks [42].

While certainly considerable progress has been made, this progress was limited mostly to discrete adaptive networks, where the dynamical states of both nodes and links are chosen from a small set of discrete values. By contrast continuous adaptive networks where links and/or nodes carry a continuous valued state remain relatively unexplored. Extending our understanding of adaptive networks to continuous systems is an important goal because it may facilitate the transfer of insights into applications, where variables are often continuous. Furthermore, as we show in Chap. 4, the continuous framework may ultimately enable a deeper understanding of some phenomena. In the present book we present a first humble approaches to the mathematical analysis of continuous adaptive networks. The discussion of the dynamics of these networks profits immensely from a framework of concepts defined by statistical physics, whereas most of the actual analysis are carried out using a combination of tools from nonlinear dynamics and network science.

The central difficulty that has to be overcome in any mathematical analysis of adaptive networks is dealing with a huge-dimensional dynamical system. A detailed micro-level description of an adaptive networks must account for the presence or absence of all potential links in the system, which means that the number of variables in such detailed models typically scales with the square of the number of nodes. Even for adaptive networks with only a hundred nodes, which is small in the context of almost all applications, the corresponding dynamical system contains typically on the order of 10.000 variables, which poses a considerable challenge even for the most sophisticated tool of dynamical systems theory.

Present approaches solve the problem by describing the network by coarse-grained variables hence effectively reducing it to a low-dimensional system [42–54]. In these approaches a dynamical system is constructed which characterizes the state of the network by following the abundances of certain subgraphs. For continuous adaptive networks coarse graining is considerably more difficult and no reliable coarse-graining schemes have been proposed to date.

Our inability of reducing the complexity of continuous adaptive networks by coarse graining seems to be the central obstacle that has slowed progress in this area. In the present work we circumvent this obstacle by following an alternative route. Instead of using ideas from network science to coarse grain and thus reduce the dimensionality of the dynamical system, we use network methods to scale up analytical approaches of dynamical systems theory to make them applicable to high-dimensional systems.

The present monograph essentially consists of three illustrative examples that attack questions from different application areas. We start with the spontaneous synchronization of coupled oscillators, second, we address the diversification of an initially homogeneous population into different node classes, and third, the topological self-organization of adaptive networks toward a dynamically critical state. The investigation of the phenomena goes hand in hand with the investigation of the relation between local and global scales, and structural and dynamical properties and thus constitutes a direct description of emergence and self-organization. In particular, we ask which topological structures support a specific global dynamical state, which topological structures evolve from a given set of local dynamical rules, and which local rules generate a specific global behavior.

We start in Chapter 1 with a short introduction of the concepts and tool used in this work. The chapter focuses on topics from dynamical systems theory and graph theory. In addition, we introduce the concept of a phase transition, which we contrast against the concept of a bifurcation in order to employ both angles for the analysis in Chapter 4.

In Chapter 2, we study the interplay between structure and dynamics in a network of coupled phase oscillators described by the paradigmatic Kuramoto model. Here, the proposed approach can pinpoint specific defects precluding synchronization. Deriving a topological interpretation of Jacobi's signature criterion, we show that synchronization can only be achieved if the coupling network obeys specific topological conditions. These conditions do not only pertain to the topology of the complete network, but also to its topological building blocks. We can thus explore the impact of particular mesoscale structures on the stability of collective dynamical states.

In Chapter 3, we study the emergence of social structure in a population of self-interested agents. Here, our approach allows for studying the established continuous snowdrift game in a multi-agent setting. We propose a model that accounts for the ability of agents to maintain different levels of cooperation with different self-chosen partners. All agents continuously, selectively, and independently adapt the amount of resources allocated to each of their collaborations in order to maximize the obtained payoff, thereby shaping the social network. We show that the symmetries of the local dynamical rules scale up and are imprinted in non-obvious symmetries in the evolving global structure. The self-organized global symmetries imply a high degree of social coordination, while at the same time causing the emergence of privileged topological positions, thus diversifying the initially homogeneous population into different social classes.

In Chapter 4, we study a class of adaptive network models that evolve toward a topological configuration, in which the dynamics on the network become critical. We discuss how the emergence of self-organized criticality (SOC) is linked to the adaptive feedback loop, and argue that in a number of models displaying SOC this feedback is implemented according to a certain pattern. Our approach allows to determine how, and under which generic conditions the pattern generates SOC. The conceptual understanding enables us on the one hand to relate details of the setup of exemplary models to particular functions within the self-organization process. On the other hand, it allows us to formulate a generic recipe for the construction of adaptation rules that give rise to SOC. We demonstrate its applicability by constructing an adaptive Kuramoto model that self-organizes toward the onset of collective synchronized behavior.

Finally, we summarize our results in Chapter 5. We emphasize that they can feed back to both, the fundamental understanding of biological systems as well as the innovative design of technological applications. Beside the results on the studied phenomena, we discuss general aspects of the approach itself, including possible extensions of the explored methodology.

Acknowledgements

The ideas that we present in this book were born during the time that we have both worked at the Max Planck Institute for the Physics of Complex Systems in Dresden. We are grateful for the inspiring time we had, and for the interesting people we met and collaborated with. In particular, we would like to thank the people that were involved in the projects discussed here: Lars Rudolf, Stefano Boccaletti, Felix Droste, Jeremias Epperlein, and Stefan Siegmund.

List of abbreviations

JSC   Jacobi's signature criterion
BCC   bidirectionally connected component
nBCC   non-bidirectionally connected component
IP   inflection point
SOC   self-organized criticality

1

Concepts and tools

The purpose of this work is to develop new analytical approaches that capture the dual nature of dynamical networks by combining tools from graph theory, dynamical systems theory, and statistical physics. In this chapter we introduce the relevant concepts and approaches. The selected topics are elementary, and can be found in most textbooks. They are presented here to illustrate the line of reasoning employed, rather than to achieve mathematical rigor. In the same spirit, lengthy or technical aspects, which substantiate the argumentation but interrupt the line of thought, are presented in boxes throughout the subsequent chapters.

We begin in Section 1.1 with dynamical systems. Following a brief definition of the central notions, we focus on two techniques, linear stability analysis and bifurcation analysis, which will prove essential in the course of the following chapters. In Section 1.2, we introduce the concept of phase transitions and contrast it against the concept of a bifurcation. Finally, in Section 1.3 we review the relevant notions from graph and network theory.

1.1 Dynamical systems

In many different disciplines, observations of an experimental system are described by a set of state variables. The common reading in mathematics and physics interprets the values of these variables at any instance t as coordinates in the abstract space of all possible states of the system, the so-called *phase space* [8, 55, 56]. The instantaneous state of a system is thus described by a position in phase space. And its evolution is given by a trajectory through phase space.

Dynamical models mimic the progression of a physical systems through phase space by formulating a prescription which, for any point in phase space, specifies the points that the system will pass through in the immediate future. Typically, such prescriptions are formulated as differential equations or time discrete maps, which can then be studied in the framework of dynamical systems theory. The evolution equations may be stochastic or deterministic. Often, they are *non-linear* and parameter-dependent. If they do not explicitly depend on time, the dynamical system is called *autonomous* [56].

In this work, we consider dynamical systems described by autonomous systems of ordinary differential equations (ODEs):

$$\frac{d}{dt}\mathbf{x}(t) = \mathbf{f}(\mathbf{x}(t),\ p_1,\ldots,p_m)\ , \tag{1.1}$$

where the components of the vector field $\mathbf{f}$ are smooth functions depending only on the phase-space coordinates $\mathbf{x} \in \mathbb{R}^n$, and on m parameters $p_i \in \mathbb{R}$ [8]. A function $\mathbf{x}(t)$ which solves the system of ODEs for a given set of initial conditions is called a *trajectory* or *orbit*. While in many textbooks the emphasis is on the calculation and the characterization of individual trajectories (see [57,58] and references therein), this work mostly focusses on families of trajectories, and, in particular, on their long-term behavior [8,56].

The dynamical systems studied in this work are *dissipative*. This means that over time the phase space volume spanned by trajectories with different initial conditions contracts. In such systems, we can distinguish between transient and long-term behavior. During the transients, the system approaches certain regions in phase space, in which it then remains for all time.

One possible type of long-term behavior is stationarity. It corresponds to a steady state $\mathbf{x}^*$ of the system, for which $\mathbf{f}(\mathbf{x}^*) = 0$. Below, we employ the example of a steady state to discuss the concept of stability, and the techniques of linear stability analysis.

The notion of stability as used in this work can be understood in terms of the system's reaction to small perturbations. A steady state $\mathbf{x}^*$ of the dynamical system given by Eq. (1.1) is called stable, if all trajectories beginning close to it remain close – that is, if small perturbations from the steady state remain small. The stronger notion of local asymptotic stability requires that all trajectories starting close to the steady state eventually converge toward it, i.e., that small perturbations decay. Below, we use the term 'stability' in the sense of local asymptotic stability.

In the vicinity of a steady state, the system of ODEs in Eq. (1.1) can be approximated by the linear system

$$\frac{d}{dt}\delta\mathbf{x} = \mathbf{J}\,\delta\mathbf{x}\,. \tag{1.2}$$

Here, $\delta\mathbf{x}$ denotes a small deflection from $\mathbf{x}^*$ and $\mathbf{J} \in \mathbb{R}^{n\times n}$ is the *Jacobian matrix*, defined by $J_{ik} = \partial\dot{x}_i/\partial x_k|_{\mathbf{x}^*}$. The solutions of the linear system are well known, and can be written in the form

$$\delta\mathbf{x}(t) = \sum_{i=1}^{n} c_i \, e^{\lambda_i t} \, \mathbf{v}_i \,, \tag{1.3}$$

where the c_i are coefficients predefined by the initial conditions, $\mathbf{v}_i$ are the right eigenvectors of $\mathbf{J}$ and λ_i the corresponding eigenvalues [56]. Equation (1.3) shows that the stability of $\mathbf{x}^*$ is determined by the spectrum of $\mathbf{J}$: if all eigenvalues of the Jacobian have negative real parts, then all the perturbations $\delta\mathbf{x}$ decay, and the steady state is stable. By contrast, if one or more eigenvalues of the Jacobian have positive real parts, then perturbations in direction of the respective eigenvectors are amplified, and the steady state is unstable.

In summary, the linear stability analysis of a dynamical system reduces to an eigenvalue problem. If all parameters p_i are specified, the problem can be solved by using numerical algorithms for calculating the spectrum of high-dimensional matrices [59]. In many cases, however, it is interesting to study families of systems with generic, unspecified parameters. In such cases, the eigenvalue problem has to be solved analytically. In particular for high-dimensional systems, it is useful to consider that for determining the stability properties, complete knowledge of the spectrum of $\mathbf{J}$ is unnecessary; knowledge of the signs alone suffices. Tools for stability analysis that depart from this point are the method of resultants [60], and *Jacobi's signature criterion* discussed in Chapters 2 and 3.

The idea of studying families of dynamical systems, whose members differ with respect to their parameter values p_i, leads directly to another important concept, that of a *bifurcation*. In general, two systems with slightly different parameter sets $\mathbf{p}$ and $\mathbf{p}'$ have different but *topologically equivalent* dynamics, that is, they are identical with respect to the number, and type of their attractors. However, there may be parameter sets $\mathbf{p}$, for which there are arbitrarily close sets $\mathbf{p}'$ exhibiting topologically non-equivalent dynamics. Such parameter sets $\mathbf{p}$ define special points in *parameter space*, called bifurcation points [8, 56]. Crossing a bifurcation point in parameter space induces a structural transition in the phase space. The transition itself is called a bifurcation.

If the evolution equations (1.1) are smooth functions, the bifurcation points are not isolated in parameter space, but lie on a manifold [56]. The difference between the dimension of the parameter space and the dimension of the bifurcation manifold is referred to as the *codimension* of the bifurcation. In practise, the codimension can be understood as the minimal number of parameters that need to be varied for the bifurcation to occur.

Bifurcation manifolds divide the parameter space in regions of topologically equivalent systems. In physics, such regions are often called the *phases* of a system. Phases and bifurcations represent the observable characteristics of a dynamical system, which can be used to compare the theoretical model with the described experimental system. Thus, variations of experimental parameters within a phase should not qualitatively change the behavior of the experimental system, while already marginal changes close to a bifurcation manifold are expected to induce qualitative transitions [61–63].

1.2 Phases and phase transitions

Many particle systems in equilibrium can often be analyzed on a macroscopic, phenomenological level. The classical example is the ideal gas: Statistical physics relates the microscopic properties of individual particles, such as location and momentum of atoms and molecules, to macroscopic state variables, such as temperature and pressure. The relation between both levels of description is probabilistic. It is based on the assumption that the probability of finding a state with given macroscopic state variables depends on the number of compatible microscopic states [64].

Generalizing statistical physics to systems far from equilibrium, considered here, is still a major challenge. We thus do not use statistical physics for the analysis of the systems that we consider in the subsequent chapters. Yet, we do use its terminology to describe the featured phenomena on a macroscopic, phenomenological level.

In the language of statistical physics, a phase is a region in parameter space, throughout which the macroscopic state variables of a system change smoothly. At a phase boundary, by contrast, small variations of the parameters induce qualitative changes in the macroscopic variables. These changes can either be discontinuous, or else be continuous, but non-analytic. Depending on the actual type of changes, we speak of a discontinuous/continuous *phase transition* respectively [65, 66].

In this work, we will be concerned only with continuous phase transitions. A system that undergoes such a transitions displays unique features, which are commonly subsumed under the term *criticality*. In a critical state, the correlation between local properties of the microscopic constituents extend over arbitrarily large distances, limited only by the size of the system [66]. Moreover, a number of observables show power-law behavior.

Continuous phase transitions can in many cases be studied in terms of a so-called *order parameter*, an observable that it is zero in one phase and non-zero in the other [66]. An *order parameter profile* associates to every point in parameter space a value of the order parameter. It displays kinks at the points in parameter space, where the transition occurs. It is worth noticing that specifying an order parameter determines the corresponding phase transition, while a phase transition can often be described by different order parameters [65, 67].

In summary, a situation, in which small variations in the parameters induce qualitative changes in the behavior of an experimental system, can be analyzed in terms of a bifurcation or a phase transition depending the description that is available for the underlying system. In cases where both description are available, the relation between bifurcations and phase transitions is known [68].

For the class of systems described in Chapter 4, the phase-transition treatment is well established; descriptions in terms of bifurcation analysis, however, are so far known only for a few special cases [69]. Yet, even for systems which have no such descriptions, insights from bifurcation theory can often be used indirectly.

1.3 Graphs

In the following chapters, we will use the tools from dynamical systems to analyze the microscopic dynamics of and on networks, and the concepts from statistical physics to describe the observed macroscopic behavior. The concepts of graph theory will be used to classify the networks' structure.

Let us emphasize that there exist two terminological frameworks for describing the structure of a network: That of graph theory, which it mostly used in mathematics, and that of complex network theory, which is more common in physics. Here, we mainly use the terminology of complex network theory. The only exception are the terms 'network', and 'network theory', which we replace by the terms 'graph', and 'graph theory', where we want to make explicit that solely structural aspects are meant.

Below, we give a short overview over the central notions of complex network/graph theory. Detailed introductions into the field can be found in [70, 71].

A graph describes the pairwise relations between objects from a certain collection. The objects are represented by so-called *nodes*; their pairwise relations by so-called *links*. Every link has two endpoints in the collection of nodes, and is said to connect them. Connected nodes are called *neighboring* or *adjacent*, the aggregate of all neighborhood relations defines the *topology*.

Graphs are categorized in different classes depending on the properties of their links [71]. A link that has an orientation, distinguishing one of the connected nodes as its origin and the other as its end point, is called *directed*. Equivalently, a link that has no orientation is called *undirected*. A link that is associated with a real number is called *weighted*. A directed/undirected/weighted graph is a graph that contains directed/undirected/weighted links.

For describing the structure of a graph, one often refers to its building blocks. A graph $\mathcal{G}'$, whose nodes and links form subsets of the nodes and link of a given graph $\mathcal{G}$, is called a *subgraph* of $\mathcal{G}$. A subgraph $\mathcal{G}'$ of $\mathcal{G}$ is called a spanning subgraph if it contains all nodes of $\mathcal{G}$.

A (simple) *path* is a subgraph that consists of a set of nodes connected by an unbranched sequence of links; the length of the path is the number of links contained [71]. A *cycle* is a path that ends at the node it begins. Cycles of length one are called self-loops.

A graph is called *connected* if every pair of nodes is connected by a path, and it is called *fully connected* if every pair of nodes is connected by a path of length one. A graph that is not connected can be divided into disjoint connected subgraphs, called *components*. A component without cycles is a *tree*.

The complete topology of a graph is captured in the so-called *adjacency matrix*. This is a square matrix **A**, whose order is determined by the number of nodes in the graph. An entry A_{ij} of **A** is zero if node i and j are not adjacent, and one if they are.

To capture the topology of weighted graphs one can define a *weighted adjacency matrix* **W**. This is an adjacency matrix, whose entries W_{ij} denote the weight of the link between i and j. Thus, W_{ij} is zero if node i and j are not adjacent, but can usually take any value from a specified interval if they are.

Some of the global attributes of graphs defined above map directly to global properties of the adjacency matrix. Thus, for an undirected graph, $A_{ij} = A_{ji}$, i.e., the adjacency matrix is symmetric. For a graph without self-loops, it has zeroes on the diagonal. Finally, if a graph consists of more than one component, its node can be relabeled such that the adjacency matrix becomes block diagonal [71].

In network models that contain a large number of links and nodes, the structure of the underlying graph is often described only on the level of global topological features. The most prominent among these features are based on the *degree* of the nodes. A nodes's degree is defined as the number of links attached to it [70]. The *degree distribution* of a graph captures the relative frequency of nodes with a given degree. The *mean degree* of a graph is the average of the degrees of its nodes and thus proportional to the ratio between links and nodes of the graph. Interestingly, this coarse grained measure already allows some statements concerning the actual topology. Thus, a component with N nodes and mean degree $2(N - 1)/N$ must be a tree, a component with N nodes and a mean degree greater or equal to 2 must have at least one cycle.

Finally, note that the locality of interactions and information, which is fundamental to our definition of self-organization, is topological in nature. In the context of graphs and networks, locality is usually defined as pertaining to direct neighborhood relations. For instance, in Chapter 4 we call an update rule local if it only depends on the state of a focal node and the states of its topological neighbors.

1.4 Summary

In this chapter we have reviewed tools and concepts from dynamical systems theory, statistical physics, and graph theory, which are used for the analysis of self-organization phenomena in network models in the subsequent chapters. In these models, the evolution of the individual microscopic constituents are mostly formulated in terms of dynamical systems. However, the dimension of the systems is typically huge, such that the emergent, system-level phenomena fall within the scope of the concepts of many particle statistical physics. Finally, the structure of the networks will be described using the concepts from graph theory.

2

Topological stability criteria for synchronized states

In the multifarious world of emergent collective phenomena, synchronization stands out as one of the most intensively studied [72, 73]. The term pertains to situations, in which coupled microscopic oscillators acquire a common frequency due to their interactions. This then gives rise to macroscopic oscillations.

One reason for the prominent role of synchronization is the universality of the phenomenon. Real world examples of systems, whose periodic macroscopic behavior can be traced back to collective oscillations of individual microscopic units, are found in many different fields such as biology, ecology, and engineering [72–75]. They include for instance the beating heart [76], populations of flashing fireflies [77] or radio communication devices [78].

Another reason for the extensive study of synchronization is that it is considered a benchmark for the understanding of emergent phenomena [72]. The paradigmatic model proposed by Kuramoto [79] has opened the field for detailed studies of the interplay between collective dynamics and interaction structure [80–84]. These studies have revealed the influence of various topological measures, such as the clustering coefficient, the diameter, and the degree or weight distribution, on the propensity to synchronize [85–87]. However, recent results [74, 88, 89] indicate that beside global topological measures also details of the exact local configuration can crucially affect synchronization. This highlights synchronization of phase oscillators as a promising example, in which it may be possible to understand the local, global, and mesoscale constraints on stability that severely limit the operation of complex technical and institutional systems [90, 91].

In this chapter, we derive necessary topological conditions for the stability of collective synchronized dynamics. The presentation follows a series of original publications in which this approach was developed [92–94]. While previous statistical approaches have reveal the influence of global topological

9

features on the propensity to synchronize, the method described here identifies specific topological defects precluding synchronization. These pertain to subgraphs that contain multiple nodes but are smaller than the entire network, thus imposing constraints on the mesoscale.

We start in Section 2.1 by introducing the Kuramoto-model of heterogeneous coupled phase oscillators. By means of Jacobi's signature criterion we determine necessary algebraic conditions for the stability of phase-locked solutions. In Section 2.2, we introduce a graphical notation, on the basis of which the algebraic stability conditions can be mapped to topological stability criteria (cf. Section 2.3). In Section 2.4, we show that the results can readily be generalized to other systems than the studied. Finally, we apply our analysis to an adaptive version of the Kuramoto model in Section 2.5, and discuss our results in Section 2.6.

2.1 Stability in networks of phase-oscillators

Consider a system of N oscillators i, whose time evolution is given by

$$\dot{x}_i = \omega_i + \sum_{j \neq i} A_{ij} \sin(x_j - x_i) , \quad \forall i \in 1 \ldots N . \tag{2.1}$$

Here, x_i and ω_i denote the phase and the intrinsic frequency of node i, while $\mathbf{A} \in \mathbb{R}^{N \times N}$ is the weight matrix of an undirected, weighted interaction network.

Equation (2.1) defines the so-called Kuramoto model, that is today considered to be a paradigm for the study of synchronization phenomena in coupled discrete systems [95], and is, therefore, used as the natural benchmark for comparative evaluations of performances of methods and tools.

If $A_{ij} \neq 0$, the two oscillators i and j are said to be coupled, if further $\dot{x}_i(t) = \dot{x}_j(t)$ for all t, they are said to be *phase locked*. In this chapter, we are interested in completely phase locked states, i.e., in states where $\dot{x}_i(t) = \dot{x}_j(t)$ for all i, j. When studied in a reference frame that co-rotates with a frequency $\Omega = 1/N \sum_{i=1}^{N} \omega_i$, the phase-locked states correspond to steady states of the governing system of Eqs. (2.1). The local stability of such states is determined by the eigenvalues of the Jacobian matrix $\mathbf{J} \in \mathbb{R}^{N \times N}$ defined by $J_{ik} = \partial \dot{x}_i / \partial x_k$ (cf. Chapter 1).

In systems of symmetrically coupled phase oscillators, the Jacobian $\mathbf{J}$ is symmetric and thus admits analysis by Jacobi's signature criterion (JSC). The

JSC (also known as Sylvester criterion) states that a hermitian or symmetric matrix **J** with rank r has r negative eigenvalues if and only if all principal minors of order $q \leq r$ have the sign of $(-1)^q$ [96]. Here, the principal minor of order q is defined as $D_q := \det(J_{ik})$, $i, k = s_1, \ldots, s_q$.

Stability analysis by means of JSC is well-known in control theory [96] and has been applied to problems of different fields from fluid- and thermo-dynamics to offshore engineering [97–99]. However, the applicability of JSC is presently limited to systems with few degrees of freedom. For system with many degrees of freedom the analytical evaluation of JSC is impeded by the growth of both, a) the number of determinants that have to be checked, and b) the number of terms in each determinant. Dealing with this growth is the central difficulty addressed in the present chapter.

Let us first consider difficulty (a) stated above. Applying the sufficient condition is impracticable for most larger systems. Note, however, that demanding $\text{sgn}(D_q) = (-1)^q$ for some q already yields a necessary condition for stability.

The necessary stability condition that is found by considering a principal minor of given order q depends on the ordering of variables, i.e., the ordering of rows and columns in the Jacobian. By considering different orderings, the number of conditions obtained for a given q can therefore be increased [100]. To distinguish minors that are based on different orderings of the variables, we define $S = \{s_1, \ldots, s_q\}$ as a set of q indices and $D_{q,S}$ as the determinant of the submatrix of **J**, which is spanned by the variables $x_{s_1}, \ldots, x_{s_q}$. Therewith, the conditions for stability read

$$\text{sgn}(D_{q,S}) = (-1)^q, \quad \forall S, \ q = 1, \ldots, r. \tag{2.2}$$

Considering necessary rather than sufficient conditions avoids the difficulty (a) mentioned above, which leaves us to deal with difficulty (b), i.e., the combinatorial explosion of terms that are needed to write out the conditions for increasing q. In the common notation more than 700 terms are necessary for expressing the minors of order 6. For circumventing this problem we employ a graphical notation that captures basic intuition and allows for expressing the minors in a concise way.

2.2 Graphical notation

We propose a graphical notation based on a topological reading of the minors [92]. We interpret the Jacobian **J** as the weight matrix of an undirected, weighted

---

The Leibniz formula for determinants [101] implies that (i) a minor $D_{q,S}$ is a sum over $q!$ elementary products $J_{i_1 j_1} \cdot \ldots \cdot J_{i_q j_q}$; and that (ii) in each of these products every index $s_i \in S$ occurs exactly twice.

In the topological reading this translates to the following statements: Because of property (i), each term of a minor $D_{q,S}$ corresponds to a subgraph with q links. Because of property (ii), these subgraphs are composed of sets of cycles in $\mathcal{G}$: Every index $s_i \in S$ occurs either with multiplicity two on a diagonal element of $\mathbf{J}$, or, with multiplicity one, on two off-diagonal elements of $\mathbf{J}$. In the former case, the respective factor corresponds to a self-loop of $\mathcal{G}$, i.e., to a cycle of length $n = 1$; in the latter case, there is a set of factors J_{ij} $i \neq j$ corresponding to a closed path of links, i.e., a cycle of length $n > 1$.

Box 1 All minors $D_{q,S}$ can be decomposed in the cyclic subgraphs of $\mathcal{G}$.

graph $\mathcal{G}$. A Jacobian element J_{ij} then corresponds to the weight of a link connecting nodes i and j. We can now relate products of the Jacobian elements to subgraphs of $\mathcal{G}$ spanned by the respective links. For instance $J_{ij} J_{jk}$ is interpreted as the path i-j-k, $J_{ij} J_{jk} J_{ki}$ as a cycle from i to j to k and back to i. Thus, the minors of $\mathbf{J}$ can be expressed as sums over subgraphs of $\mathcal{G}$.

In Box 1 we show that every term occurring in a minor $D_{q,S}$ of $\mathbf{J}$ corresponds to a subgraph that can be decomposed in cycles of $\mathcal{G}$. This allows expressing the index structure of every term by a combination of symbols denoting cycles of a given length. The idea is now to supplement the basis of symbols with a summation convention; This convention is designed such that all algebraic terms that are structurally identical and only differ by index permutations can be captured in one symbolic term, which drastically reduces the complexity of the minors.

Below, we use the following definitions: The basis of symbols is given by $\times, |, \triangle, \square, \ldots$ denoting cycles of length $n = 1, 2, 3, 4, \ldots$. The summation convention stipulates that in a minor $D_{q,S}$, every product of symbols denotes the sum over all non-equivalent possibilities to build the depicted subgraph with the q nodes $s_1, \ldots, s_q$. With these conventions the first 4 principal minors can be written as

$$D_{1.S} = \times \tag{2.3a}$$
$$D_{2.S} = \times \cdot \times - | \tag{2.3b}$$
$$D_{3.S} = \times \cdot \times \cdot \times - \times \cdot | + 2\triangle \tag{2.3c}$$
$$D_{4.S} = \times \cdot \times \cdot \times \cdot \times - \times \cdot \times \cdot | + | \cdot | + 2 \times \cdot \triangle - 2\square \tag{2.3d}$$

More generally

$$D_{q.S} = \sum \left(\text{all combinations of symbols with } \sum n = q\right), \tag{2.4}$$

where symbols with $n > 2$ appear with a factor of 2 that reflects the two possible orientations in which the corresponding subgraphs can be paced out. Symbols with an even (odd) number of links carry a negative (positive) sign related to the sign of the respective index permutation in the Leibniz formula for determinants [101].

An example for the graphical notation is presented in Figure 2.1. The figure displays the three principal minors of the symmetric 3×3 matrix

$$\mathbf{J} = \begin{pmatrix} J_{11} & J_{12} & J_{13} \\ J_{12} & J_{22} & J_{23} \\ J_{13} & J_{23} & J_{33} \end{pmatrix} \tag{2.5}$$

in algebraic, and graphical notation. Moreover, it displays for each term the corresponding subgraph of a three-node graph $\mathcal{G}$.

In many systems, including the standard Kuramoto model, fundamental conservation laws impose a zero-row-sum condition, such that $J_{ii} = -\sum_{j \neq i} J_{ij}$. Using this relation we can remove all occurrences of elements J_{ii} from the Jacobian and its minors. In the topological reading this substitution changes the graph $\mathcal{G}$ by replacing a self-loop at a node i by the negative sum over all links that connect to i.

The simplification of the minors due to the zero-row-sum condition can be understood using the example of Eqs. (2.3). Replacing the self-loops, the first term of every minor $D_{q.S}$, $\times^q$, is $(-1)^q$ times the sum over all subgraphs meeting the following conditions: (i) the subgraph contains exactly q links, and (ii) it can be drawn by starting each link at a different node in S. By means of elementary combinatorics it can be verified that all other terms of $D_{q.S}$ cancel exactly those subgraphs in $\times^q$ that contain cycles. This enables us to express the minors in another way: Defining

$$\Phi_{q.S} = \sum \text{all acyclic subgraphs of } \mathcal{G} \text{ meeting conditions (i) and (ii)} \tag{2.6}$$

$$D_{1,S} = J_{11}$$

$$= \times$$

$$D_{2,S} = J_{11}J_{22} - J_{12}^2$$

$$= \times \cdot \times - |$$

$$D_{3,S} = J_{11}J_{22}J_{33} - (J_{11}J_{23}^2 + J_{22}J_{13}^2 + J_{33}J_{12}^2)$$

$$+ 2 \cdot J_{12}J_{23}J_{13}$$

$$= \times \cdot \times \cdot \times - \times \cdot | + 2 \cdot \triangle$$

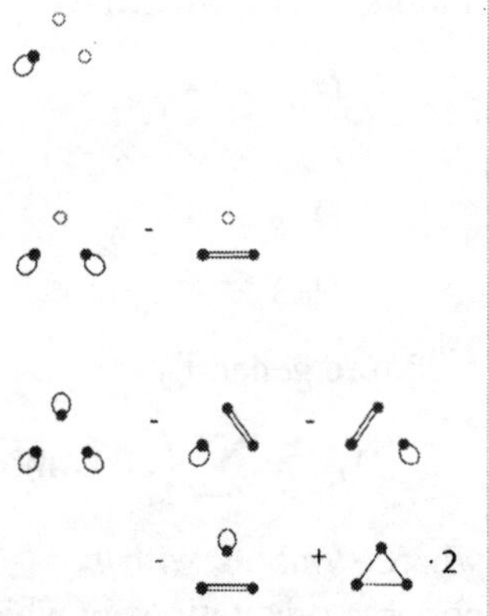

Figure 2.1 Example for the graphical notation. Shown are the minors of the matrix (2.5) in algebraic, and graphical notation, and, for each term, the corresponding subgraph of a three-node graph $\mathcal{G}$. Here, as well as in the next figures, filled symbols correspond to nodes $\in S$, open symbols to nodes $\notin S$.

we can write

$$D_{q,S} = (-1)^q \Phi_{q,S} . \tag{2.7}$$

We remark that Kirchhoff's theorem [102], which has previously been used for the analysis of dynamical systems [103, 104], appears as the special case of Eq. (2.7), in which $q = N - 1$.

The simplification of the minors due to the zero-row-sum condition as well the relation between the $D_{q,S}$ and their topological equivalents $\Phi_{q,S}$ can be illustrated by means of a simple example. Consider the symmetric 6×6 matrix

$$\mathbf{J} = \begin{pmatrix} -(J_{12}+J_{13}) & J_{12} & J_{13} & 0 & 0 & 0 \\ J_{12} & -(J_{12}+J_{23}) & J_{23} & 0 & 0 & 0 \\ J_{13} & J_{23} & -(J_{13}+J_{23}+J_{34}) & J_{34} & 0 & 0 \\ 0 & 0 & J_{34} & -(J_{34}+J_{45}) & J_{45} & 0 \\ 0 & 0 & 0 & J_{45} & -(J_{45}+J_{56}) & J_{56} \\ 0 & 0 & 0 & 0 & J_{56} & -J_{56} \end{pmatrix} \tag{2.8}$$

which obeys the zero-row-sum condition. In Figure 2.2, we calculate the minor $D_{4,S=\{1,...,4\}}$ in terms of the subgraphs of the corresponding graph $\mathcal{G}$. The calculation illustrates the reasoning that lead to Eqs. (2.6) and (2.7).

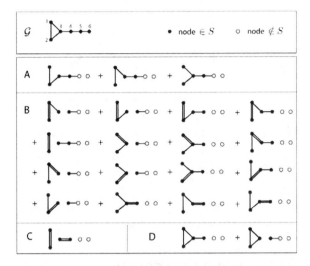

Figure 2.2 Symbolic calculation of a minor using the zero-row-sum condition. Shown is the graph $\mathcal{G}$, defined by the off-diagonal entries of Eq. (2.8). The terms of the minor $D_{4,S}$ can be written as $\times \cdot \times \cdot \times \cdot \times = A + B + C + 2D, - \times \cdot \times \cdot | = -(B+2C), | \cdot | = C, 2 \times \cdot \triangle = -2D$ and $-2\square = 0$ (cf. Eq. (2.3d)). It thus follows that $D_{4,S} \equiv \Phi_{4,S} = A$ is the sum over all acyclic subgraphs of $\mathcal{G}$ meeting conditions (i) and (ii).

The complete sequence of minors $D_{q,S}, q = 1, \ldots, r$ can be calculated as

$$D_{1,S} = (-1)(J_{12} + J_{13})$$
$$D_{2,S} = (-1)^2 (J_{12}J_{13} + J_{12}J_{23} + J_{13}J_{23})$$
$$D_{3,S} = (-1)^3 (J_{12}J_{13} + J_{12}J_{23} + J_{13}J_{23}) J_{34}$$
$$D_{4,S} = (-1)^4 (J_{12}J_{13} + J_{12}J_{23} + J_{13}J_{23}) J_{34}J_{45}$$
$$D_{5,S} = (-1)^5 (J_{12}J_{13} + J_{12}J_{23} + J_{13}J_{23}) J_{34}J_{45}J_{56}$$

where $S = \{1, \ldots, q\}$, and $r = 5$ due to the zero-row-sum condition.

On the other hand, we can use the definition (2.6) to construct the sequence $\Phi_{q,S}, q = 1, \ldots, r$, directly from the graph $\mathcal{G}$ (cf. Figure 2.3). A comparison of both, the algebraic and the topological results, reproduces Eq. (2.7). Note that labeling the nodes in different order would have yielded different algebraic as well as topological expressions.

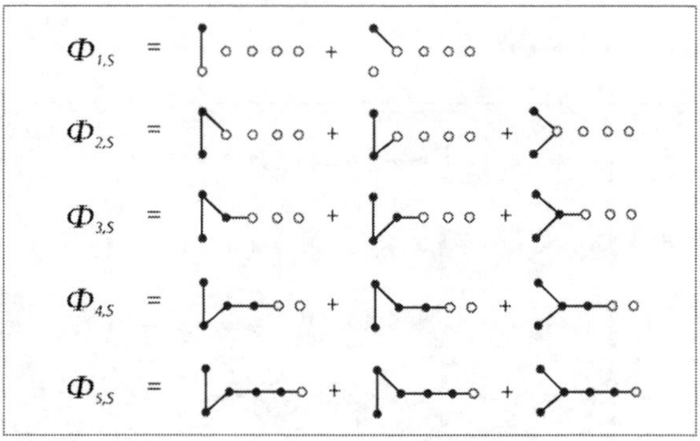

Figure 2.3 Topological equivalents of minors. Shown is the complete sequence of $\Phi_{q,S}, q = 1, \ldots, 5$ for the graph $\mathcal{G}$ from Figure 2.2.

2.3 Topological stability conditions

Let us shortly summarize what we obtained so far. The topological reading of determinants maps a symmetric Jacobian $\mathbf{J}$ with zero row sum onto a graph $\mathcal{G}$, whose weighted adjacency matrix is given by the off-diagonal part of $\mathbf{J}$. The minors of $\mathbf{J}$ can then be interpreted as sums over values associated with subgraphs of $\mathcal{G}$. Combining Eqs. (2.2) and (2.7), the algebraic stability constraints on the minors of $\mathbf{J}$ translate into

$$\Phi_{q,S} > 0, \quad \forall S, \; q = 1, \ldots, r. \tag{2.9}$$

We emphasize that the graph $\mathcal{G}$ is not an abstract construction, but combines information about the physical interaction topology and the dynamical units. For example, if a graph $\mathcal{G}$ has disconnected components, there is a reordering of the variables x_i, such that $\mathbf{J}$ is block diagonal. This implies that the spectra of different topological components of $\mathcal{G}$ decouple and can thus be treated independently.

From Eq. (2.9) we can immediately read off a weak *sufficient* condition for stability: As $\Phi_{q,S}$ is a sum over products of the J_{ij}, a Jacobian with $J_{ij} \geq 0$ $\forall i, j$ is a solution to Eq. (2.9) irrespective of the specific structure of $\mathcal{G}$ (for the specific case of the Kuramoto model known as Ermentrout theorem [105, 106]). By contrast, if $J_{ij} < 0$ for some i, j, then the existence of solutions of Eq. (2.9) is dependent on the topology.

The Φ-notation allows to investigate which combinations of negative J_{ij} in a graph $\mathcal{G}$ lead to the violation of at least one of Eqs. (2.9). For this

Let $X \subset I_1$, be the subset of nodes $\in I_1$ that are incident to at least one link from E^* ('boundary nodes'). Let x_i denote the elements of X; and let σ_m denote the sum over all elements of E^* incident to a node x_m. Let further A, B and C denote subsets of X and $\sigma_A := \prod_{m \in A} \sigma_m$. Per definition, $(-1)^{|A|}\sigma_A > 0$ for all $A \neq \emptyset$. Finally, let τ_Y, $Y \subset X$ be the sum over all woods of $\mathcal{G}$ that (i) span I_1, and (ii) consists of $|Y|$ trees each of which contains exactly one element from Y. For the sake of clarity, we use the abbreviation $\Phi_S := \Phi_{|S|,S}$.

The redefinition of Φ reveals that

$$\Phi_{I_1 \backslash C} = \sum_{B \subseteq X \backslash C} \sigma_B \tau_{B \cup C} . \tag{2.10}$$

We can thus write

$$\sum_{C \subseteq X} \underbrace{(-1)^{|C|} \sigma_C}_{>0} \underbrace{\Phi_{I_1 \backslash C}}_{>0} = \sum_{C \subseteq X} (-1)^{|C|} \sigma_C \sum_{B \subseteq X \backslash C} \sigma_B \tau_{B \cup C}$$

$$= \sum_{\substack{C \subseteq X \\ B \subseteq X \backslash C}} (-1)^{|C|} \sigma_{B \cup C} \tau_{B \cup C}$$

$$= \sum_{\substack{A \subseteq X \\ C \subseteq A}} (-1)^{|C|} \sigma_A \tau_A$$

$$= \sum_{A \subseteq X} \sigma_A \tau_A \underbrace{\sum_{C \subseteq A} (-1)^{|C|}}_{=0} = 0$$

which is a contradiction.

Box 2 Spanning tree criterion for stability: Proof sketch

purpose, a definition of Φ, which is equivalent to the one given above, proves advantageous: $\Phi_{N-x,S}$ is the sum over all decompositions of $\mathcal{G}$ in x components, each of which is either an isolated node $\notin S$ or a tree with exactly one node $\notin S$.

In Box. 2, we show that Eq. (2.9) requires that every component of $\mathcal{G}$ has a spanning tree, whose links all have positive weights. The sketched proof uses the redefinition of Φ to show that at least one of the necessary

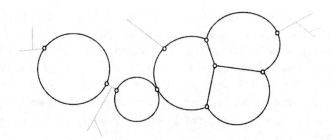

Figure 2.4 Decomposition of a graph $\mathcal{G}$ in acyclic parts (grey lines), unbranched segments of cycles (black lines) and branching points (open circles). Lines represent paths of $\mathcal{G}$ that may contain an arbitrary number of nodes. Circles represent single nodes. Stability requires that (i) the acyclic parts of $\mathcal{G}$ only contain links with positive weights; and that (ii) any unbranched segment of a cycle of $\mathcal{G}$ contains at most one link with negative weight.

conditions is violated when no spanning tree of positive connections exists. It thereby also provides an illustration of the direct application of the notation in a calculation.

We start by defining $I = \{1, \dots, N\}$ as the set of all node labels i. If $\mathcal{G}$ does not have a spanning tree, whose links all have positive weights, it exists a partition of $I = I_1 \cup I_2$ such that

$$J_{ij} \leq 0 \quad \forall\, i, j \mid i \in I_1,\; j \in I_2 \,. \tag{2.11}$$

Equivalently, all elements $(E^*)_i$, of the set $E^* := E(I_1) \cap E(I_2)$ are negative.

The key idea is now to formally expand a well defined series of $\Phi_{q.s}$ in the contributions of I_1, I_2, and E^*. It can be then be shown that all of these $\Phi_{q.s}$ being positive is incompatible with all $(E^*)_i$ being negative. Let us emphasize that the sketched approach dispenses with the specific calculation of any $\Phi_{q.s}$, and hence with addressing the up to $q!$ terms they subsume.

In addition to the spanning tree criterion for stability, which pertains to a global property of $\mathcal{G}$, Eq. (2.9) implies further restrictions, which pertain to mesoscale properties of $\mathcal{G}$. Thus, the weights of all links that are not part of any cycle of $\mathcal{G}$ have to be positive. Moreover, at most one of the links that build an unbranched segment of a cycle of $\mathcal{G}$ may have a negative weight (cf. Figure 2.4). Finally, the weight of such a 'negative link' is bounded below by a value that depends on the weights of the other links in the segment (cf. Box 2).

Consider a path of $d - 1$ degree-two nodes v_i, $i \in \{1, \ldots, d - 1\}$, that are part of at least one cycle of $\mathcal{G}$. Together with the d links c_i that are incident to at least one of the nodes v_i, the path constitutes an unbranched segment C of a cycle of $\mathcal{G}$ (cf. Figure 2.4).

According to the spanning tree criterion, C can have at most one link with negative weight. Below, we consider the case where C has exactly one such link, and show that Eq. (2.9) imposes an upper bound on the absolute value of the negative link weight.

We use following conventions: Let the nodes be labeled such that the indices i occur in an increasing order if C is paced out. And let the links be labeled such that v_i is incident to c_i and c_{i+1}. Further, let w_i denote the weight of c_i. And lastly, let c_x denote the only link in C with negative weight w_x. Below, show that stability requires that

$$|w_x| < \frac{\prod \text{all } w_i, \, i \in I^*}{\sum \text{all distinct products of } (d - 2) \text{ factors } w_i, \, i \in I^*}, \quad (2.12)$$

where $I^* = \{1, \ldots, d\} \setminus x$.

For deriving Eq. (2.12), we consider a sequence of conditions $\Phi_{S_1} \ldots \Phi_{S_{d-1}}$. The sequence is constructed as follows: we choose $S_1 = x$, $S_2 = \{x, x + 1\}$, $S_3 = \{x, x + 1, x + 2\}$, and so forth until $S_{d-x} = \{x, x + 1, x + 2, \ldots, d - 1\}$. The remaining elements of the sequence are then constructed as $S_{d-x+1} = S_{d-x} \cup \{x - 1\}$, $S_{d-x+2} = S_{d-x} \cup \{x - 1, x - 2\}$, and so forth until $S_{d-1} = \{v_i\}_{i \in \{1, \ldots, d-1\}}$.

The first element of the sequence, $\Phi_{S_1} > 0$, stipulates that $w_x + w_{x+1} > 0$, and thus that $-w_x < w_{x+1}$.

The second element of the sequence $\Phi_{S_2} > 0$ stipulates that $w_x w_{x+1} + w_x w_{x+2} + w_{x+1} w_{x+2} > 0$ and thus that $-w_x < (w_{x+1} w_{x+2})/(w_{x+1} + w_{x+2})$.

More generally, every element Φ_{S_i} of the sequence $\Phi_{S_1}, \ldots, \Phi_{S_{d-1}}$ modifies the upper bound on $-w_x$ as per

$$-w_x < \frac{\prod \text{all } w_i, \, i \in S_i^*}{\sum \text{all distinct products of } (l(S_i) - 1) \text{ factors } w_i, \, i \in S_i^*}, \quad (2.13)$$

where $S_i* = (S_i \setminus x) \cup (\max(S_i) + 1)$.

The right hand side of Eq. (2.13) is monotonously decreasing with increasing $l(S_i)$. We can thus conclude that C can have at most one link, whose absolute value $|w_x|$ is bounded above by Eq. (2.12).

Box 3 Lower bound on negative link weights

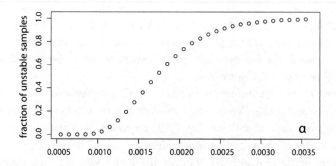

Figure 2.5 Numerical test of the hypothesized stability condition. The fraction of matrices **J** that have a positive largest eigenvalue although the corresponding graph $\mathcal{G}$ possesses a positive spanning tree is plotted against $|\alpha|$. The continuous transition from 0 to 1 confirms that the upper bound on $|\alpha|$ depends on the exact position of the negative link.

Note that while the mesoscale criteria restricting the number and position of negative links can be subsumed under the global spanning tree criterion, the mesoscale criteria restricting the weights of possible negative links are inherently bound to the mesoscale. In particular, the latter can only be derived by considering $\Phi_{q,s}$ with small and intermediate q. This highlights the benefits of the proposed approach: The JSC provide stability criteria on all scales q, which are made accessible by means of the graphical notation and symbolic calculus.

In order to check the derived topological stability criteria numerically, we generated ensembles of 10^8 connected graphs of size $N = 25$ and fixed mean degree $\langle k \rangle = 4$. In each graph, we assigned a negative weight α to all but $N - 1$ randomly chosen links. The remaining links were assigned weight 1. We then checked for each graph, whether the graph had a positive spanning tree and calculated the largest non-trivial eigenvalue λ of the corresponding Jacobian. The procedure was repeated for different values of α.

Among the 10^9 generated test graphs, more than 98% did not contain a positive spanning tree. Of these networks none were found to be stable which complies with the spanning tree criterion for stability. Among the graphs that did contain a positive spanning tree (ca. 10^7), stability depended on the specific topology and the value of $|\alpha|$ (cf. Figure 2.5). As expected from Eq. (2.12), the fraction of networks that are unstable although they obey the spanning tree criterion increases with increasing $|\alpha|$.

Applied to the Kuramoto model defined in Eq. (2.1), the topological stability criteria reveal common properties of the configuration of all possible phase-locked systems. In a phase-locked state, $J_{ij} = A_{ij} \cos(x_j - x_i)$, where $x_j - x_i$ is the stationary phase difference between oscillator j and i. Given that all link weights $A_{ij} \geq 0$, stability of the phase-locked state requires that the coupling network has a spanning tree of oscillators obeying $|x_j - x_i| < \pi/2$.

2.4 Other applications

Although we have so far focused on the Kuramoto model as an example, the topological interpretation of the JSC only requires a hermitian Jacobian. Further, the simplification leading up to the spanning tree criterion is possible whenever the Jacobian has zero row sums. This condition is satisfied for instance by all systems of the form

$$\dot{x}_i = C_i + \sum_{j \neq i} A_{ij} \cdot O_{ij}(x_j - x_i), \quad \forall i \in 1 \dots N \tag{2.14}$$

where the A_{ij} are the weights of a symmetric interaction network and the O_{ij} odd functions. We emphasize that the approach remains applicable in heterogeneous networks containing different link strengths A_{ij}, coupling functions O_{ij} or intrinsic parameters C_i.

The class of systems, to which the present results are directly applicable, thus includes general networks of phase oscillators as well as other models such as continuous-time variants of the Deffuant model of opinion formation [107]. In the context of this model, the x_i denote opinions held by a networked community of agents. The criteria derived here then constrain the distribution of opinions that can be sustained in stationarity.

Although the specific criteria derived above are contingent on the zero-row sum condition, it can be expected that the general approach proposed here is also applicable to situations where this condition is violated, such as in the model of cooperation among interacting agents studied in [100]. A simple extension of Eq. (2.14), which violates the zero row sum condition, is found by replacing C_i by a function of x_i. Such an extension can account for the dynamical retuning of the intrinsic frequency, e.g., for modeling homeostatic feedback in neural networks. For illustrating the application of the proposed method to models, in which the zero row sum condition is violated in some rows, we consider another model inspired by neuroscience in the subsequent section.

Before we address our final example, let us emphasize that the graphical notation proposed here may also be useful for problems not concerning stability analysis. For instance, it allows exploring the isospectrality of hermitian or symmetric matrices [108, 109], which differ with respect to the signs of some off-diagonal entries: The characteristic polynomial χ of a hermitian matrix $\mathbf{A} \in \mathbb{C}^{n \times n}$ can be expressed as $\chi(\lambda) = D_n(\mathbf{A} - \lambda \mathbf{I})$. Considering the structure of the graph $\mathcal{G}$ associated to $\mathbf{A} - \lambda \mathbf{I}$, allows to determine which symbols contribute to χ. On this basis, we can then address the isospectrality of matrices that are generated from $\mathbf{A}$ by a mapping $R : A_{ij} \to -A_{ij}, \ A_{ji} \to -A_{ji}$, which changes the sign of an arbitrary number of off-diagonal entries while preserving hermiticity. A map R that leaves all symbols in $D_n(\mathbf{A} - \lambda \mathbf{I})$ invariant leaves the characteristic polynomial and thus the spectrum invariant. That is, we may evaluate the global impact of an operation R by only regarding its effect on the mesoscale structures, to which it directly contributes.

As an example, consider a hermitian matrix $\mathbf{A} - \lambda \mathbf{I}$, whose corresponding graph $\mathcal{G}$ is a tree. The only symbols that contribute to the characteristic polynomial χ are the symbols $\times$ and $|$. Symbols $\times$ correspond to factors $(A_{ii} - \lambda)$ in χ, symbols $|$ to factors $A_{ij}A_{ji} = |A_{ij}|^2$. Both are invariant under R. Hence, the spectrum of matrices $\mathbf{A}$, whose corresponding graphs $\mathcal{G}$ is a tree, is invariant under any operation that changes the sign of a pair of off-diagonal entries.

Along the same line, we can infer isospectrality relations for matrices $\mathbf{A}$, whose corresponding graphs $\mathcal{G}$ are composed only of tree-like subgraphs and isolated cycles. For such matrices, the spectrum is invariant under hermiticity preserving sign changes of

1. an arbitrary number of off-diagonal entries that do not belong to cyclic subgraphs of $\mathcal{G}$.
2. an even number of off-diagonal entries that belong to the same cyclic subgraph of $\mathcal{G}$.

2.5 Adaptive Kuramoto model

We now apply the proposed approach to an example, in which the topology of the interaction network coevolves with the dynamics of the oscillators [17, 20, 37]. In the context of the Kuramoto model, adaptive coupling has recently attracted keen interest as it allowed to show that the appearance of synchronous motion can be intimately related to a selection mechanism of specific network topologies [110], and to the identification of complex hierarchical structures in the graph connectivity [37, 111].

We consider a system of N phase oscillators that evolve according to Eq. (2.1), while the coupling strength A_{ij} evolves according to

$$\frac{d}{dt} A_{ij} = \cos(x_j - x_i) - b \cdot A_{ij} . \qquad (2.15)$$

The first term in Eq. (2.15) states that the more similar the phases of two nodes the stronger reinforced is their connection, the second term guarantees convergence. In a stationary, phase-locked state state, $A_{ij} = \cos(x_j - x_i)/b$ and all oscillators oscillate with a common frequency $\Omega = \frac{1}{N}\sum_i \omega_i$. The stability of this state is governed by a symmetric Jacobian

$$\mathbf{J} = \left(\begin{array}{ccc|ccc}
-b & 0 & 0 & s_{21} & s_{12} & 0 \\
0 & -b & 0 & s_{31} & 0 & s_{13} \\
0 & 0 & -b & 0 & s_{32} & s_{23} \\
\hline
s_{21} & s_{31} & 0 & m_1 & o_{12} & o_{13} \\
s_{12} & 0 & s_{32} & o_{12} & m_2 & o_{23} \\
0 & s_{13} & s_{23} & o_{13} & o_{23} & m_3
\end{array}\right), \qquad (2.16)$$

where

$$o_{ij} := \tfrac{1}{b}\cos^2(x_j - x_i), \quad m_i := -\sum_{j\neq i} o_{ij}, \quad s_{ji} := \sin(x_j - x_i),$$

and we have chosen $N = 3$ for illustration. The marked partitioning separates two blocks on the diagonal. The upper one is a diagonal submatrix of size $L \times L$, $L := N(N-1)/2$, the lower one is a $N \times N$ symmetric submatrix with zero row sum, which we denote as $\mathbf{j}$.

Let us start our analysis by focusing on the upper left block of $\mathbf{J}$. In the chosen ordering of variables, the first L minors $D_{|S|.S}$ satisfy the stability condition (2.2) iff $b > 0$. Concerning the minors of order $|S| > L$, the following conventions prove advantageous: Below, we consider sets S that contain all variables A_{ij} and n of the variables x_i. For every such set, we define S' as the subset of S that only contains the n variables x_i. Moreover, we define $\tilde{\mathbf{j}}$ as the matrix that is obtained from $\mathbf{j}$ if all o_{ij} are substituted by $\cos(2(x_j - x_i))$. Finally, we define $\tilde{D}_{n.S'}$ as the determinant of the submatrix of $\tilde{\mathbf{j}}$, which is spanned by the variables $\in S'$.

We find that

$$D_{L+n,S} = (-1)^L b^{L-n} \cdot \tilde{D}_{n.S'}. \qquad (2.17)$$

As $\tilde{\mathbf{j}}$ is symmetric and has a zero row sum, its minors, $\tilde{D}_{n,S}$, can be rewritten using Eq. (2.7)

$$D_{L+n,S} = (-1)^{L+n} b^{L-n} \tilde{\Phi}_{n,S'}, \qquad (2.18)$$

where $\tilde{\Phi}_{n,S'}$ refers to subgraphs of the graph $\tilde{\mathcal{G}}$ defined by the off-diagonal entries of $\tilde{\mathbf{j}}$. Stability requires that $\mathrm{sgn}\left(D_{L+n,S}\right) = \mathrm{sgn}\left((-1)^{L+n}\right)$. As the necessary stability condition $b > 0$ implies $b^{L-n} > 0$, it follows that in a stable system

$$\tilde{\Phi}_{n,S'} > 0, \quad \forall S, \; n = 1, \ldots, \mathrm{rank}(\tilde{\mathbf{j}}). \tag{2.19}$$

Comparison with Eq. (2.9) reveals that a necessary condition for stability is that every component of the graph $\tilde{\mathcal{G}}$ has a positive spanning tree. Revisiting the definition of graph $\tilde{\mathcal{G}}$, we find that the weight of a link ij of $\tilde{\mathcal{G}}$ is given by $\cos(2(x_j - x_i))$. Hence, every component of $\tilde{\mathcal{G}}$ has to have a positive spanning tree iff every component of the adaptive coupling network has a spanning tree of oscillators obeying $|x_j - x_i| < \pi/4$. The restriction on the stationary phase-differences in a stable, phase-locked state are thus more strict in the adaptive than in the non-adaptive case.

2.6 Discussion

In this chapter, we analyzed necessary conditions for local asymptotic stability of stationary and phase-locked states in networks of phase oscillators.

Using a graphical interpretation of Jacobi's signature criterion we showed that for all non-trivial eigenvalues of $\mathbf{J}$ to be negative, the graph $\mathcal{G}$, whose adjacency matrix is given by the off-diagonal part of $\mathbf{J}$, has to obey necessary topological conditions. Thus, $\mathcal{G}$ must have a spanning tree of links with positive weights, which restricts the number and position of potential negative entries in the Jacobian matrix. Moreover, the absolute value of potential negative link weights is bounded above by a topology depended relation.

Our results provide an analytical angle that is complementary to statistical analysis of network synchronizability. Where statistical approaches reveal global features impinging on the propensity to synchronize, our approach can pinpoint specific defects precluding synchronization. We note that such defects can occur on all scales, corresponding to the violation of the signature criterion in subgraphs of different size. This highlights synchronization of phase oscillators as a simple but intriguing example in which instabilities can arise from local, global or mesoscale structures. In the future, the approach proposed here may provide a basis for further investigation of these instabilities.

Testing the conditions identified here in real-world systems requires information on the stationary phase profile. This limits the applicability of our approach for synchronizing systems that do not synchronize naturally. However, we note that even in such systems it may be possible to stabilize an existing unstable phase-locked state, e.g. by delayed-feedback control [112]. Based on the observed phase profile in the stabilized state, one can then use the identified conditions to search for structures that preclude synchronization when the controller is turned off. A more direct application of the present results is possible in networks with designed phase profiles [113]. Given a coupling topology and a desired phase profile, it is often relatively easy to find a set of natural frequencies for which the phase profile is stationary but not necessarily stable. Here, the necessary stability conditions provide constrains on the stable profiles that may be realized in a given coupling topology.

The topological stability conditions can directly be applied to other coupled dynamical systems. This even holds if the underlying dynamical equations are unknown as long as the Jacobian is accessible, for instance from data analysis [114] or generalized modeling [115]. Even if these specific techniques are not applied, researchers are often aware of the structure of the Jacobian in the system under consideration, which can be sufficient for gaining some fundamental insights by the proposed approach.

The present results demonstrate the applicability of Jacobi's signature criterion to large networks. In principle, the criterion can be applied to all systems in which the Jacobian is a hermitian matrix. In the present chapter, we additionally assumed that the Jacobian has zero row sums. An example of the application of Jacobi's signature criterion in a large system, in which the zero row sum condition is violated, is presented in the next chapter.

3

Patterns of cooperation

In this chapter we continue to explore the interplay of structure and dynamics in a complex, self-organizing system. However, in contrast to the abstract example studied previously we now consider a simple social model. Although its applicability is not obvious from the beginning, the methodology developed in the previous chapter will be essential for explaining the dynamics in this system.

Social structure and social dynamics are commonly studied using the example of cooperation networks. Cooperation builds the basis for many institutions that shape our lives on different scales [116, 117]: Thus, humans cooperate in communities, companies, ethnies and nations [118–120]. Thereby, the collaborative behavior of the individuals is strongly influenced by the embedding social structure, while the social structure itself evolves in response to the individuals' collaborative behavior [121]. The adaptive interplay is implicated in a variety of emergent phenomena [122, 123]. In this chapter, we investigate its role for the emergence of social coordination, diversification, and for the rise of leaders that hold distinguished social positions.

The interplay between collaborative behavior and social structure has first been explored in evolutionary game theory [124]. The intense research in this field has identified several mechanism allowing for the evolution and persistence of cooperation despite the often high costs incurred by an cooperating agent [125]. In particular, the emergence of cooperation is promoted if the interacting agents are distributed in some (potentially abstract) space, so that only certain agents can interact at any given time [126–128]. In the context of social cooperation, spatial structure can be appropriately modeled by a complex network, in which nodes represent agents, while the links correspond to collaborations. The topology of this network has been shown to be of central importance for the level of cooperation that evolves [129–132].

In social networks, the topology is not static, but reacts to the behavior of the agents [133–138]. Adaptive network models, which account for the dynamical interplay, have been studied for some time in the social literature (e.g [138–140]), while pioneering work [141, 142] only recently triggered a wave of detailed dynamical investigations in physics [143]. Recent publications discuss simple cooperative games such as the one-shot prisoner's dilemma [144–154], the iterated prisoner's dilemma [155,156], and the snow-drift game [21, 157, 158] on adaptive networks. They showed numerically and analytically that a significantly increased level of cooperation can be achieved if individuals are able rewire their links [21, 144–149, 159, 160], if links are formed and broken [150, 151, 157–159, 161], or if new agents are added to the network [153, 154]. Moreover, it has been shown that the adaptive interplay between the agents' strategies and the network topology can lead to the emergence of distinguished agents from an initially homogeneous population [142, 144–146].

While important progress has been made in the investigation of games on adaptive networks, it is mostly limited to discrete networks, in which the agents can only assume a small number of different states, say, unconditional cooperation with all neighbors or unconditional defection. Most current models therefore neglect the ability of intelligent agents to maintain different levels of cooperation with different self-chosen partners [162].

In this chapter, we analyze a weighted and directed adaptive network model, in which agents continuously and selectively reinforce advantageous collaborations. The material presented here combines two previous publications [100,163]. After a brief description of the model, we show in Section 3.3 that the network generally approaches a state in which all agents make the same total cooperative investment and every reciprocated investment yields the same benefit. Despite the emergence of this high degree of coordination, the evolved networks are far from homogeneous. Typically, the agents distribute their total investment heterogeneously among their collaborations, and each collaborations receives different investments from the partners. In Section 3.5, we show that this heterogeneity enables resource fluxes across the network, which allow agents holding distinguished topological positions to extract high payoffs. Thereafter, in Section 3.6, we investigate further topological properties of the evolved networks and identify the transition in which large cooperating components are formed. In Section 3.7, we then focus on the appearance of unidirectional (unreciprocated) investments. Specifically, we identify three distinct scenarios in which unidirectional collaborations can arise and discuss their implications for the interaction topology. Finally, in Section 3.8, we discuss a possible extension of the model which accounts for a player's success feeding back on his cooperative investment Our conclusions are summarized in Section 3.9.

3.1 The model

We consider a population of N agents, representing for instance people, firms or nations, engaged in bilateral collaborative interactions. Each interaction is described by a continuous snowdrift game [117], one of the fundamental models of game theory. In this game, an agent i can invest an amount of time/money/effort $e_{ij} \in \mathbb{R}_0^+$ into the collaboration with another agent j. Cooperative investments accrue equal benefits B to both partners, but create a cost C for the investing agent. Assuming that investments from both agents contribute additively to the creation of the benefit, the payoff received by agent i from an interaction with an agent j can then be written as

$$P_{ij} = B \left(e_{ij} + e_{ji} \right) - C \left(e_{ij} \right) . \tag{3.1}$$

The game thus describes the generic situation, in which agents invest their personal resources to create a common good shared with the partner.

As an example of the snowdrift game, the reader may think of a scientific collaboration where two researchers invest their personal time in a project, while the benefit of the publication is shared between them. This example makes clear that the benefit of the collaboration must saturate when an extensive amount of effort is invested, whereas the costs faced by an agent, measured for instance in terms of personal well-being, clearly grows superlinearly once the personal investment exceeds some hours per day.

In the following, we do not restrict the cost- and the benefit-functions, B and C, to specific functional forms, except in the numerical investigations. However, we assume that both are differentiable and, moreover, that B is sigmoidal and C is superlinear (cf. Figure 3.3). These assumptions capture basic features of real-world systems such as inefficiency of small investments, saturation of benefits at high investments, as well as additional costs incurred by overexertion of personal resources and are widely used in the sociological and economic literature [164, 165].

To account for multiple collaborations per agent, we assume that the benefits received from collaborations add linearly, whereas the costs are a function of the sum of investments made by an agent, such that the total payoff received by an agent i is given by

$$P_i = \sum_{j \neq i} B\left(\sigma_{ij}\right) - C\left(\Sigma_i\right) . \tag{3.2}$$

where $\Sigma_i := \sum_{j=1}^{N} e_{ij}$ denotes the *total investment* of the agent i while $\sigma_{ij} := e_{ij} + e_{ji}$ denotes the total investment made in the collaboration ij. This is motivated by considering that benefits from different collaborations, say different publications, are often obtained independently of each other, whereas the costs generated by different collaborations stress the same pool of personal resources.

Let us emphasize that we do not restrict the investment of an agent further. While investments cannot be negative, no upper limit on the investments is imposed. Furthermore, the agents are free to make different investments in collaborations with different partners. Thus, to optimize its payoff, an agent can reallocate investments among its potential partners as well as change the total amount of resources invested.

For specifying the dynamics of the network, we assume the agents to be selfish, trying to increase their total payoff P_i by a downhill-gradient optimization

$$\frac{\mathrm{d}}{\mathrm{d}t} e_{ij} = \frac{\partial}{\partial e_{ij}} P_i . \tag{3.3}$$

Every agent can cooperate with every other agent. Thus, the network of potential collaborations is fully connected and the deterministic time-evolution of the model system is given by a system of $N(N-1)$ ordinary differential equations of the form of Eq. (3.3). The network dynamics, considered in the following, is therefore only the shifting of link weights e_{ij}. Note, however, that already the weight dynamics constitutes a topological change. As will be shown in the following, the agents typically reduce their investment in the majority of potential collaborations to zero, so that a sparse and sometimes disconnected network of non-vanishing collaborations is formed. Therefore the terminology of graph theory is useful for characterizing the state that the system approaches. Below, we use the term *link* to denote only those collaborations that receive a non-vanishing investment σ_{ij}. A link is said to be *bidirectional* if non-vanishing investments are contributed by both connected agents, while it is said to be *unidirectional* if one agent makes a non-vanishing investment without reciprocation by the partner. Likewise, we use the term *neighbors* to denote those agents that are connected to a focal agent by non-vanishing collaborations and the term *degree* to denote the number of non-vanishing collaborations, in which a focal agent participates.

3.2 Numerical investigation

In the following, the properties of the model are investigated mostly by analytical computations that do not require further specifications. Only for the purpose of verification and illustration we resort to numerical integration of the ODE system. For these we use the functions

$$B\left(\sigma_{ij}\right) = \frac{2\rho}{\sqrt{\tau + \rho^2}} + \frac{2(\sigma_{ij} - \rho)}{\sqrt{\tau + \left(\sigma_{ij} - \rho\right)^2}} \, , \quad C\left(\Sigma_i\right) = \mu\left(\Sigma_i\right)^2 .$$

For studying the time-evolution of exemplary model realizations by numerical integration, all variables e_{ij} are assigned random initial values drawn independently from a Gaussian distribution with expectation value 1 and standard deviation 10^{-14}, constituting a homogeneous state plus small fluctuations. The system of differential equations is then integrated using Euler's method with variable step size h. In every timestep, h is chosen such that no variable is reduced by more than half of its value in the step. If in a given timestep a variable e_{ij} falls below a threshold $\epsilon \ll 1$ and the corresponding time derivative is negative, then de_{ij}/dt is set to zero for one step to avoid very small time steps. We emphasize that introducing the threshold ϵ is done purely to speed up numerical integration and does not affect the results or their interpretation. In particular, we confirmed numerically that, the exact value of ϵ does not influence the final configuration that is approached. In all numerical results shown below, $\epsilon = 10^{-5}$ was used.

3.3 Coordination of investments

The numerical exploration of the system reveals frustrated, glass-like behavior; starting from a homogeneous configuration as described above, it approaches either one of a large number of different final configurations, which are local maxima of the total payoff.

A representative example of an evolved network, and snapshots from the time-evolution of two smaller example networks are shown in Figures 3.1 and 3.2, respectively. In the example networks only those links are shown that receive a non-vanishing (i.e. above-threshold) investment. Most of these

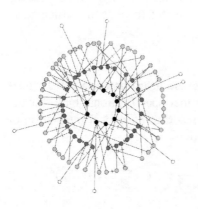

Figure 3.1 Network of collaborations in the final state. The nodes represent agents, links correspond to collaborations receiving non-vanishing investments σ_{ij}. The small dash on every link ij is a fairness indicator: the further it is shifted toward one agent i, the lower the fraction, e_{ij}/σ_{ij}, of the investment agent i contributes to the link. Agents extracting more payoff are shown in darker colour and are placed toward the center of the community. The size of a dot indicates the agent's total investment Σ_i. In the final configuration, the network exhibits a high degree of heterogeneity. Nevertheless, all agents make the same total investment and all collaborations receive the same total investment. (Parameters: $\rho = 0.65$, $\tau = 0.1$, $\mu = 1.5$)

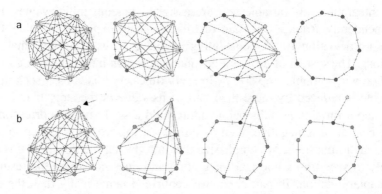

Figure 3.2 Time evolution of example networks from a homogeneous state. The different frames show snapshots of the network of collaborations at different times. a) In small systems the network sometimes self-organizes to homogeneous topologies in which all players extract the same payoff. b) If a player (arrow) tries to maintain too many links at too low investment, his partners will cease reciprocating investments, leading sometimes to unidirectional links.

non-vanishing links are *bidirectional*, receiving investments from both of the agents they connect. Only rarely, *unidirectional* links appear, which are maintained by one agent without reciprocation by the partner.

For further investigations, it is useful to define a *bidirectionally connected component* (BCC) as a set of agents and the bidirectional links connecting them, such that, starting from one agent in the set, every other agent in the set can be reached by following a sequence of bidirectional links. In the numerical investigations, we observe that all bidirectional links within a BCC receive the same total investment in the final state. However, the investment σ_{ij} made in every given link is in general not split equally among the two connected agents. Furthermore, all agents within a BCC make the same total cooperative investment Σ_i in the final state. However, the investments e_{ij} of one agent in different collaborations are in general different. The *coordination* of total investments σ_{ij}, Σ_i therefore arises although no agent has sufficient information to compute the total investment made by any other agent. We emphasize that the level of investments, which the agents approach is not set rigidly by external constraints, but instead depends on the topology of the network of collaborations that is formed dynamically. This is evident for instance in differences of up to 20% between the level of investment that is reached in different BCCs of the same network.

To understand how coordination of investment arises, we now formalize the observations made above. We claim that in our model in the final state the following holds: Within a BCC (i) every agent makes the same total investment, and (ii) either all bidirectional links receive the same total investment or there are exactly two different levels of total investment received by bidirectional links. For reasons described below, the case of two different levels of total investment per link is only very rarely encountered. In this case, every agent can have at most one bidirectional link that is maintained at the lower level of investment.

We first focus on property (i). This property is a direct consequence of the stationarity of the final state. Consider a single link ij. Since both investments, e_{ij} and e_{ji}, enter symmetrically into σ_{ij}, the derivative of the benefit with respect to either investment is $\partial B(\sigma_{ij})/\partial e_{ij} = \partial B(\sigma_{ji})/\partial e_{ji} =: B'(\sigma_{ij})$. Thus, if $e_{ij}, e_{ji} > 0$, the stationarity conditions $de_{ij}/dt = de_{ji}/dt = 0$ require

$$\frac{\partial}{\partial e_{ij}} C\left(\Sigma_i\right) = B'\left(\sigma_{ij}\right) = \frac{\partial}{\partial e_{ji}} C\left(\Sigma_j\right). \qquad (3.4)$$

Equation (3.4) implies $C'(\Sigma_i) = C'(\Sigma_j)$. As we assumed C to be superlinear, C' is injective and it follows that $\Sigma_i = \Sigma_j =: \Sigma$, such that i and j, are at a point of identical total investment. Iterating this argument along a sequence of bidirectional links yields (i).

Let us remark that the stationarity of vanishing investments may be fixed due to the external constraint that investments have to remain non-negative. The stationarity condition for vanishing and uni-directional links, analogous to Eq. (3.4), is therefore

$$\frac{\partial}{\partial e_{ij}} C\left(\Sigma_i\right) \geq B'\left(\sigma_{ij}\right) \leq \frac{\partial}{\partial e_{ji}} C\left(\Sigma_j\right). \tag{3.5}$$

Because of the inequalities that appear in this equation, the argument given above does not restrict the levels of total investment found in different components. For similar reasons agents that are only connected by unidirectional links can sustain different levels of investment, which is discussed in Section 3.7.

We note that, although the network of potential interactions is fully connected, no information is transfered along vanishing links. Therefore, the equation of motion, Eq. 3.3, should be considered as a local update rule, in the sense that it only depends on the state of the focal agent and on investments received from a small number of direct neighbors.

In order to understand property (ii), we consider multiple links connecting to a single agent i. In a steady state, the investment into each of the links has to be such that the slope of the benefit function of each link is identical. Otherwise, the payoff could be increased by shifting investments from one link to the other. Since the benefit function is sigmoidal, a given slope can be found in at most two points along the curve: one above and one below the inflection point (IP). By iteration, this implies that if a stationary level of investment is observed in one link, then the investment of all other links of the same BCC is restricted to one of two values, which amounts to the first sentence of (ii). For understanding why the case of two different levels of investments is rarely encountered, the stability of steady states has to be taken into account. A local stability analysis, based on linearisation and subsequent application of Jacobi's signature criterion, is the subject of the next section.

3.4 Stability conditions

To determine the local asymptotic stability of the steady states, we study the Jacobian matrix $\mathbf{J} \in \mathbb{R}^{N(N-1)\times N(N-1)}$ defined by $J_{(ij)(kl)} = \partial \dot{e}_{ij}/\partial e_{kl}$. The

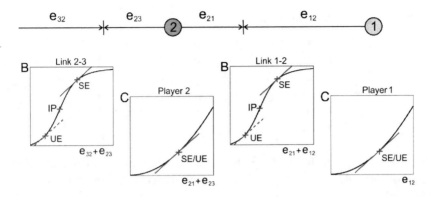

Figure 3.3 Adjustment of investments. Shown are the perceived cost functions C and benefit functions B (insets) for the example of an agent 1 of degree one interacting with an agent 2 of degree two (sketched). The function B depends on the sum of both agents' investments into the interaction while C depends on the sum of all investments of one agent. In every equilibrium (SE or UE) stationarity demands that the slope of these functions is identical. This requires that the agents make identical total investments. In stable equilibria (SE), the operating point lies in general above the inflection point (IP) of B, whereas equilibria found below the IP are in general unstable (UE). Therefore, in a stable equilibrium both links produce the same benefit and both agents make the same total investment.

terms contained in this matrix can be grouped into three different types

$$A_{ij} := \frac{\partial \dot{e}_{ij}}{\partial e_{ij}} = \frac{\partial^2}{(\partial e_{ij})^2} B\left(\sigma_{ij}\right) - \frac{\partial^2}{\left(\partial e_{ij}\right)^2} C\left(\Sigma_i\right) \qquad (3.6)$$

$$P_{ij} := \frac{\partial \dot{e}_{ij}}{\partial e_{ji}} = \frac{\partial}{\partial e_{ji}} \frac{\partial}{\partial e_{ij}} B\left(\sigma_{ij}\right) \qquad (3.7)$$

$$K_i := \frac{\partial \dot{e}_{ij}}{\partial e_{il}} = -\frac{\partial}{\partial e_{il}} \frac{\partial}{\partial e_{ij}} C\left(\Sigma_i\right) \qquad (3.8)$$

albeit evaluated at different points. For reasons of symmetry

$$\frac{\partial}{\partial e_{ji}} \frac{\partial}{\partial e_{ij}} B\left(\sigma_{ij}\right) = \frac{\partial^2}{\left(\partial e_{ij}\right)^2} B\left(\sigma_{ij}\right) =: B''\left(\sigma_{ij}\right)$$

$$\frac{\partial}{\partial e_{il}} \frac{\partial}{\partial e_{ij}} C\left(\Sigma_i\right) = \frac{\partial^2}{\left(\partial e_{ij}\right)^2} C\left(\Sigma_i\right) =: C''\left(\Sigma_i\right) ,$$

and consequentially $P_{ij} = P_{ji}$, and $A_{ij} = P_{ij} + K_i$. Ordering the variables according to the mapping $M : \mathbb{N} \times \mathbb{N} \to \mathbb{N}; \ (i, j) \to N(i - 1) + j$ the

Jacobian can be written in the form

$$
\mathbf{J} = \begin{pmatrix}
A_{12} & K_1 & P_{12} & 0 & 0 & 0 \\
K_1 & A_{13} & 0 & 0 & P_{13} & 0 \\
P_{12} & 0 & A_{21} & K_2 & 0 & 0 \\
0 & 0 & K_2 & A_{23} & 0 & P_{23} \\
0 & P_{13} & 0 & 0 & A_{31} & K_3 \\
0 & 0 & 0 & P_{23} & K_3 & A_{32}
\end{pmatrix}, \tag{3.9}
$$

which is shown here for $N = 3$. As each cooperation ij is determined by a pair of variables (e_{ij}, e_{ji}), each P_{ij} occurs twice forming quadratic sub-units with the corresponding entries A_{ij} and A_{ji}. Subsequently, we restrict ourselves to the submatrix $\mathbf{J}^s$ of $\mathbf{J}$, which only captures variables e_{ij} belonging to 'non-vanishing' links. As argued before, 'vanishing links', i.e. links with $\sigma_{ij} = 0$, are subject to stationarity condition (3.5). If $C'(\Sigma_i) > B'(0)$, their stability is due to the boundary condition $e_{ij} \geq 0$ and is independent of the second derivatives of C and B. Hence, they can be omitted from the subsequent analysis. This means in particular that the spectra of different topological components of the network decouple and can thus be treated independently.

The Jacobian $\mathbf{J}^s$ is symmetric and can hence be analyzed by means of Jacobi's signature criterion [167] introduced in Chapter 2. Thus, the system is stable if all minors D_q of $\mathbf{J}^s$ satisfy $\text{sgn}(D_q) = (-1)^q$ for $q = 1, \ldots, N(N-1)$. In chapter 2, we countered the high number of algebraic stability conditions by summarizing them in few effective topological conditions. Unfortunately, the results cannot readily be applied to the system considered here, as the Jacobian $\mathbf{J}^s$ does not obey a zero-row sum condition. However, we show below that evaluating the conditions for different minors of order $q = 1, 2$ already suffices to understand property (ii).

By means of an even number of column and row interchanges the above stated form of $\mathbf{J}^s$ can always be transformed such that the first 2×2 block reads

$$
\begin{pmatrix}
A_{ij} & P_{ij} \\
P_{ij} & A_{ji}
\end{pmatrix}.
$$

Since we assume that ij is a non-vanishing link, and, hence, i and j to be in the same component, both agents make the same total investment Σ. It follows from definition (3.8) that $K_i = K_j =: K$ and therewith that $A_{ij} =$

A_{ji}. Thus, the sequence $1, D_1, D_2$ alternates if

$$D_1 = P_{ij} + K < 0 \quad \wedge \tag{3.10}$$
$$D_2 = (2P_{ij} + K) K > 0. \tag{3.11}$$

Equation (3.11) stipulates that K and $(2P_{ij} + K)$ have the same sign. Of the two possible scenarios

$$(2P_{ij} + K), K < 0 \quad \text{and} \quad (2P_{ij} + K), K > 0 \tag{3.12}$$

the second is ruled out by Eq. (3.10): If $K > 0$, it follows from Eq. (3.10) that $P_{ij} < -K < 0$, which contradicts $(2P_{ij} + K) > 0$. Hence, the necessary conditions for stability, Eqs. (3.10), (3.11), require

$$K < 0 \quad \wedge \quad (2P_{ij} + K) < 0. \tag{3.13}$$

If either agent i or agent j has another bilateral link, say ik, it is furthermore possible to transform $\mathbf{J}^s$ by an even number of row and line interchanges such that the first 2×2 block reads

$$\begin{pmatrix} A_{ij} & K \\ K & A_{ik} \end{pmatrix}. \tag{3.14}$$

In this representation the sequence $1, D_1, D_2$ alternates if

$$D_1 = \qquad A_{ij} = P_{ij} + K < 0 \tag{3.15}$$
$$D_2 = P_{ik} P_{ij} + (P_{ik} + P_{ij}) K > 0. \tag{3.16}$$

Condition (3.16) can then be written as

$$P_{ik} P_{ij} > -K (P_{ik} + P_{ij}). \tag{3.17}$$

Inserting the definitions (3.6)–(3.8) into Eqs. (3.13) and (3.17) reveals that for a pair of agents ij connected by a bidirectional link, stability requires

$$C''(\Sigma_i) > 0 \wedge 2B''(\sigma_{ij}) - C''(\Sigma_i) < 0, \tag{3.18}$$

and every pair of links ij and ik connecting to the same agent i has to satisfy

$$B''(\sigma_{ik}) B''(\sigma_{ij}) > \underbrace{C''(\Sigma_i)}_{>0} (B''(\sigma_{ik}) + B''(\sigma_{ij})). \tag{3.19}$$

Note that Eq. (3.18) implies $2B''(\sigma_{ij}) < C''(\Sigma_i) > 0$, but does not stipulate the sign of $B''(\sigma_{ij})$. As Eq. (3.18) pertains also to the link ik, the same holds for $B''(\sigma_{ik})$. We therefore have to consider three different cases when testing the compatibility of Eq. (3.19) with Eq. (3.18):

(a) $B''(\sigma_{ik}) < 0$ and $B''(\sigma_{ij}) < 0$, (both investments above the IP)
(b) $B''(\sigma_{ik}) > 0$ and $B''(\sigma_{ij}) > 0$, (both investments below the IP)
(c) $B''(\sigma_{ik}) > 0$ and $B''(\sigma_{ij}) < 0$, (one investment above, one below the IP).

In case (a), Eq. (3.19) is trivially fulfilled as the left hand side has positive and the right hand side negative sign. In case (b), Eq. (3.19) and Eq. (3.18) are incompatible: estimating the lower bound of the right hand side of (3.19) using the relation $C''(\Sigma) > 2B''(\sigma_{ij})$ leads to the contradiction

$$
\overbrace{B''(\sigma_{ik})B''(\sigma_{ij})}^{:=X>0} > C''(\Sigma_i)\left(B''(\sigma_{ik}) + B''(\sigma_{ij})\right)
$$
$$
> 2B''(\sigma_{ij})\left(B''(\sigma_{ik}) + B''(\sigma_{ij})\right)
$$
$$
= \underbrace{2B''(\sigma_{ij})B''(\sigma_{ik})}_{=2X} + \underbrace{2\left(B''(\sigma_{ij})\right)^2}_{>0}.
$$

This shows that in a stable stationary state, every agent can at most have one link receiving investments below the IP. In case (c), Eq. (3.19) can in principle be satisfied. However, the equation still imposes a rather strong restriction on a positive $B''(\sigma_{ik})$ requiring high curvature of the benefit function close to saturation. The restriction becomes stronger, when the degree of agent i increases. This can be shown by taking determinants D_q with $q > 2$ into account.

Bilateral links with investments below the IP can be excluded entirely, if the benefit function approaches saturation softly, so that the curvature above the inflection point remains lower or equal than the maximum curvature below the inflection point. For such functions, every pair $\sigma_{ik} < \sigma_{ij}$ of solutions to the stationarity condition $B'(\sigma_{ij}) = B'(\sigma_{ik}) = C'(\Sigma_i)$ yields a pair of coefficients $B''(\sigma_{ik}) > 0$, $B''(\sigma_{ij}) < 0$ violating (3.19). In this case, only configurations in which all links receive investments above the IP can be stable and hence all links produce the same benefit in the stable stationary states. This explains why the case of two different levels of cooperation is generally not observed in numerical investigations if realistic cost and benefit functions are used.

For understanding the central role of the IP for stability, consider that in the IP the slope of B is maximal. Therefore, links close to the IP make attractive targets for investments. If the total investment into one link is below the IP, a disturbance that raises (lowers) the investment increases (decreases) the slope, thus making the link more (less) attractive for investments. Hence, below the IP, a withdrawal of resources by one of the partners, no matter how slight, will make the collaboration less attractive, causing a withdrawal by the other partner and thereby launching the interaction into a downward spiral. Conversely, for links above the IP the gradual withdrawal of resources by one partner increases the attractiveness of the collaboration and is therefore compensated by increased investment from the other partner. In psychology, both responses to withdrawal from a relationship are well known [166]. The proposed model can therefore provide a rational for their observation that does not require explicit reference to long term memory, planning, or irrational emotional attachment.

For our further analysis, property (ii) is useful as it implies that, although our model is in essence a dynamical system, the BCCs found in the steady states of this system can be analyzed with the tools of graph theory for undirected graphs. In Sections 3.5 and 3.6 we go one step further and treat not only the BCC but the whole network as an undirected graph. We thereby ignore the differences between directed and undirected links in order to study properties such as the degree- and component-size distributions before we continue in Section 3.7 with a more detailed investigation of directed links and their topological implications.

3.5 Distinguished topological positions

Despite the coordination described above, the payoff extracted by agents in the final state can differ significantly. This is remarkable because the agents follow identical rules and the network of collaborations is initially almost homogeneous with respect to degree, link weights, and neighborhood.

Because all bidirectional links in a BCC produce the same benefit, the total benefit an agent receives is proportional to the degree of the agent. By contrast, the cost incurred by an agent does not scale with the degree, but is identical for all agents in the BCC, because agents of high degree invest a proportionally smaller amount into their collaborations. Topological positions of high degree thus allow agents to extract significantly higher benefits without requiring more investment.

The payoff distribution in the population is governed by the degree distribution p_k describing the relative frequency of agents with degree k. Figure 3.4 shows a representative degree distribution of an evolved networks in the final state. While the finite width of the distribution indicates heterogeneity, the distribution is narrower, and therefore fairer, than that of an Erdős–Rényi random graph, which constitutes a null-model for randomly assembled network

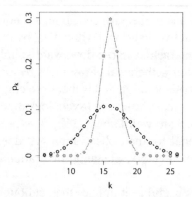

Figure 3.4 Degree heterogeneity in self-organized networks of collaborations. In comparison to a random graph (black), the degree distribution of the evolved networks is relatively narrow (grey). Parameters are chosen to obtain networks with identical mean degree. Results are averaged over 100 networks of size $N = 100$.

topologies. We verified that the variance of the evolved network is below the variance of a random graph for the whole range of admissible mean degree $\bar{k}$ in a network of given size.

Although the snowdrift game is not a zero-sum game, payoffs cannot be generated arbitrarily. In order to sustain the extraction of high payoffs by agents of high degree, investments have to be redistributed across the network. In the definition of our model, we did not include the transport of resources directly. Nevertheless, a redistribution of investments arises indirectly from the asymmetry of the agents' investments. This is illustrated in Figure 3.5. Consider for instance an agent of degree 1. This agent necessarily focuses his entire investment on a single collaboration. Therefore, the partner participating in this collaboration only needs to make a small investment to make the collaboration profitable. He is thus free to invest a large portion of his total investment into links to other agents of possibly higher degree. In this way, investments flow toward the regions of high degree where high payoffs are extracted (cf. Figure 3.5).

3.6 Formation of large components

To explore the topological properties of the networks of collaborations in the final state further, we performed an extensive series of numerical in-

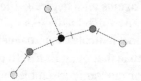

Figure 3.5 Redistribution of investments. Even in small networks investments flow toward agents of high connectivity. This flow is apparent in the position of the fairness indicators on the links, cf. Figure 3.1, caption.

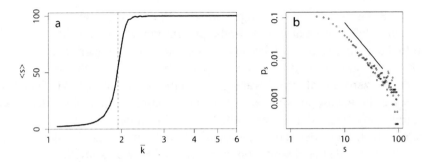

Figure 3.6 Giant component transition. (a) At $\bar{k} = 1.91$ the expected size $\langle s \rangle$ of a network component changes from $O(1)$ to $O(N)$. (b) Even in the relatively small networks of 100 nodes a power-law shape starts to appear in the component-size distribution obtained from the final states of 750 network realizations with mean degree $\bar{k} = 1.91$.

tegrations runs, in which we varied all parameters in a wide range. These revealed that an important determinant of the topology is the mean degree $\bar{k} = 2L/N$, where L denotes the number of links and N the number of agents in the network. Given two evolved networks with similar $\bar{k}$, one finds that the networks are also similar in other properties such as the component-size distribution, clustering coefficient, and the fraction of collaborations that are unidirectional. We therefore discuss the topological properties of the evolved networks as a function of $\bar{k}$, instead of the original model parameters.

We first consider the expected size $\langle s \rangle$ of a network component to which a randomly chosen agent belongs. In contrast to the BCCs discussed above, unidirectional collaborations are now taken into account in the computation of component sizes. The value of $\langle s \rangle$ in the evolved network as a function of $\bar{k}$ is shown in Figure 3.6a. The figure reveals that large components begin to appear slightly below $\bar{k} = 2$. Because of the difficulties related to integrating $N(N - 1)$ differential equations, our numerical investigations are limited to networks of up to 100 agents. While it is therefore debatable whether the observed behavior qualifies as a phase transition, it can be related to the giant component transition commonly observed in larger networks.

In the giant component transition, a component is formed that scales linearly with network size. In the absence of higher correlations, the transition occurs at $\bar{q} = 1$ [16], where $\bar{q}$ is the mean excess degree of the network, i.e., the number of additional links found connected to a agent that is reached by following a random link.

In Erdős–Rényi random graphs, $\bar{q} = \bar{k}$, therefore the giant component transition takes place at $\bar{k} = 1$. In the present model, the transition in $\langle s \rangle$ is shifted to higher values of $\bar{k}$ because of the nature of the underlying snowdrift game: The snowdrift game favors cooperation in the sense that for an agent of degree zero it is always advantageous to initiate an interaction. Therefore $\bar{k} = 1$ is the lowest possible value that can be observed in evolved networks. Further, any evolved network with $\bar{k} = 1$ invariably consists of isolated pairs, which precludes the existence of a giant component. Finally, the relatively narrow degree distribution of the evolved networks implies $\bar{q} < \bar{k}$ and therefore $\bar{k} > 1$ at the transition.

To estimate an upper limit for the connectivity at which the giant component transition occurs, it is useful to consider degree homogeneous networks. In these networks, the degree distribution is a delta function and $\bar{q} = \bar{k} - 1$, so that the transition occurs at $\bar{k} = 2$. In the networks evolved in the proposed model, we can therefore expect a critical value of $\bar{k}$ between one and two. Based on numerical results, we estimate that the giant component transition in the present model occurs at $\bar{k} \approx 1.91$ (Figure 3.6). At this value, a power-law distribution of component sizes, which is a hallmark of the giant-component transition, begins to show already in relative small networks with $N = 100$.

3.7 Unreciprocated collaborative investments

While in Section 3.3 we have mainly considered bidirectional links, and in Sections 3.5 and 3.6 only distinguished between vanishing and non-vanishing links, we will now focus on unidirectional links, which one partner maintains without reciprocation by the other. The presence of such links in collaboration networks was recently discussed in detail by Koenig et al. [168].

For the discussion below, it is advantageous to consider the mean degree of agents in a connected component $\langle k \rangle = 2l/n$, where n and l are the number of agents and links in the component. Note that in large components $\langle k \rangle \approx \bar{k}$, while the two properties can be significantly different in small components. In contrast to $\bar{k}$, $\langle k \rangle$ allows us to infer global topological properties: Components with $\langle k \rangle < 2$ are trees. Components with $\langle k \rangle = 2$ contain exactly one cycle, to which trees might be attached. And, components with $\langle k \rangle > 2$ contain more than one cycle, potentially with trees attached. As in the previous section, the term component refers to maximal subgraphs which are connected by bidirectional and/or unidirectional links. According to this definition, a component may, beside one or more BCCs, contain agents, which only have unidirectional links. In the following, we denote the set of these agents as the non-BCC part of the component (nBCC). For the sake of simplicity, we focus on components which contain only one BCC, but note that the case of multiple BCCs can be treated analogously.

Unlike the BCC, the nBCC is not a subcomponent but only a set of agents which are not necessarily connected. Nevertheless, numerical results show that (i*) all nBCC agents make the same total investment Σ_n and (ii*) all unidirectional links maintained by nBCC agents receive the same total investment σ_n. While property (ii*) can be understood analogously to property (ii) of BCCs, property (i*) cannot be ascribed to stationarity or stability conditions, but seems to result from optimality restrictions. As a consequence of the properties (i*) and (ii*), the number of outgoing links $m := \Sigma_n/\sigma_n$ is identical for all agents in the nBCC.

So far we have decomposed a component into the BCC and the nBCC. Within each subset, all agents make the same total investment, and all links receive the same total investment, therefore each subset can be characterized by two parameters, Σ_b, σ_b for the BCC and Σ_n, σ_n for the nBCC. To re-combine the subsets and infer properties of the whole component, we need to study the relation between these four parameters.

The central question guiding our exploration is why do agents not start to reciprocate the unidirectional investments. The lack of reciprocation implies that the unidirectional links are either less attractive or just as attractive as bidirectional links. We distinguish the two scenarios

(a) $B'(\sigma_b) = B'(\sigma_n)$,
(b) $B'(\sigma_b) > B'(\sigma_n)$.

In case (a) the unidirectional collaborations are as attractive as targets for investments as bidirectional collaborations. In typical networks, where all remaining links receive investments above the IP, this implies $\sigma_b = \sigma_n = \sigma$. Furthermore, in case (a) the stationarity condition, Eq. (3.4), requires that $C'(\Sigma_b) = C'(\Sigma_n)$, which stipulates $\Sigma_b = \Sigma_n =: \Sigma$. Therefore the whole component consists of agents making an investment Σ and links receiving an investment σ.

Conservation of investments within a component implies $l\sigma = n\Sigma$ and hence

$$\langle k \rangle = 2\frac{l}{n} = 2\frac{\Sigma}{\sigma} . \tag{3.20}$$

We know further that $\Sigma/\sigma = \Sigma_n/\sigma_n = m \in \mathbb{N}$, where m is the number of outgoing links of an agent in the nBCC. Inserting $\Sigma/\sigma = m$ in Eq. (3.20) yields $\langle k \rangle = 2m$, showing that unidirectional links that are as attractive as

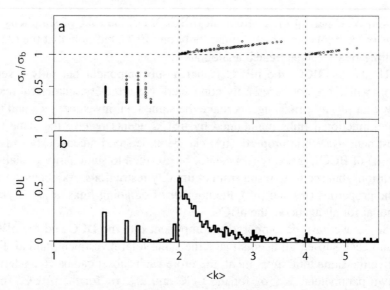

Figure 3.7 Unidirectional investments and proportion of unidirectional links. (a) The ratio between the investment in unidirectional and the investment in bidirectional links from the same component, σ_n/σ_n, equals 1 for $\langle k \rangle = 2m$, $m \in \mathbb{N}$. $\sigma_n/\sigma_b > 1$ applies to $\langle k \rangle > 2 \neq 2m$, $\sigma_n/\sigma_b < 1$ to $\langle k \rangle < 2$. (b) For $\langle k \rangle < 2$ the average proportion of unidirectional links (PUL) features discrete peaks. As every tree must have a bidirectional core, the smallest $\langle k \rangle$ with non-zero PUL is $\langle k \rangle = 4/3$. It corresponds to components with 3 agents and 2 links, one of which can be unidirectional.

bidirectional links can only occur in components, in which the mean degree $\langle k \rangle$ is an integer multiple of 2. This matches the numerical data displayed in Figure 3.7a, which shows that $\sigma_n/\sigma_b = 1$ is observed in components with $\langle k \rangle = 2$ and $\langle k \rangle = 4$.

It is remarkable that observing $\sigma_n = \sigma_b$ in a pair of collaborations is sufficient to determine the mean degree of the whole component. Moreover, components in which the mean degree is exactly 2 have to consist of a single cycle potentially with trees attached. In the numerical investigations, we mostly observe cycles of bidirectional links to which trees of unidirectional links are attached, as shown in Figure 3.8b.

In case (b) the bidirectional links are more attractive targets for investments than unidirectional links. In typical networks with $\sigma_b, \sigma_n \geq \sigma_{IP}$, this implies $\sigma_b < \sigma_n$, and thus that unidirectional links receive a higher investment than bidirectional links. Now, the stationarity condition (3.4) demands that

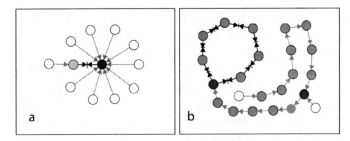

Figure 3.8 Topological arrangement of unidirectional links (depicted as grey arrows). (a) For $\langle k \rangle < 2$, unidirectional links connect individual nBCC agents with a BCC core. (b) For $\langle k \rangle \geq 2$ unidirectional links are arranged in long chains which is shown here for $\langle k \rangle = 2$. For $\langle k \rangle > 2$ typical components become too large to be presented in this way.

$C'(\Sigma_b) > C'(\Sigma_n)$, and thus that

$$\sigma_b < \sigma_n \leq \Sigma_n < \Sigma_b . \tag{3.21}$$

Hence, the total investment made by an agent investing in bidirectional links is higher than the one made by agents investing in unidirectional links. This relationship restricts the connectivity in the BCC to $\langle k \rangle_{BCC} := 2\Sigma_b/\sigma_b > 2$, which implies $\langle k \rangle > 2$, because the mean degree of the component cannot be smaller than 2 if a subcomponent already has a degree greater than 2. Therefore, we find that unidirectional links that are less attractive than bidirectional links only occur in components, in which the mean degree is larger than 2, but not an integer multiple of 2 (cf. Figure 3.7a). As such links are only found at $\bar{k}$ beyond the giant component transition, they occur typically in large components as shown in Figure 3.1.

In numerical investigations, we also observe some unidirectional links in components with $\langle k \rangle < 2$ (cf. Figure 3.7b). To explain these, we have to consider case (b) but relax the assumption that both, σ_n and σ_b are above the IP. Thus, we obtain case (c), about which we know that the unidirectional links are less attractive than bidirectional links, $\Sigma_n < \Sigma_b$, and that the unidirectional link only receives investments from one agent, i.e., $\sigma_n \leq \Sigma_n$. Moreover, $\langle k \rangle < 2$ implies $\langle k \rangle_{BCC} < 2$ and therefore $\Sigma_b < \sigma_b$. Therefore

$$\sigma_n \leq \Sigma_n < \Sigma_b < \sigma_b, \tag{3.22}$$

which shows that unidirectional links can only appear in components with $\langle k \rangle < 2$ if the investment received by unidirectional links is smaller than the investment received by bidirectional links. Satisfying simultaneously $\sigma_n < \sigma_b$ and $B'(\sigma_n) < B'(\sigma_b)$ requires $\sigma_n < \sigma_{IP}$. The components with $\langle k \rangle < 2$, in which such links are found, are trees formed by a core of bidirectional links, to which individual agents are attached by unidirectional links (Figure 3.8a). Chains of unidirectional links, as we have observed in case (a), cannot appear for $\langle k \rangle < 2$: Such a scenario would necessitate that some agents have one incoming and one outgoing link below the IP, which is ruled out by a trivial extension of the reasoning from Section 3.3.

3.8 Extension of the model

Below, we consider an extension of the studied model, in which a player's success feeds back on his cooperative investment. In reality, such feedbacks arise for instance because interest rates for loans may be lower for players, who receive a high income. Including a benefit-dependent reduction of C in the model yields a fully adaptive network.

In our adaptive model, players enjoy benefit-dependent cost reduction

$$P_{ij} = B\left(\sigma_{ij}\right) - \frac{e_{ij}}{\Sigma_i} C\left(\Sigma_i\right) \cdot \frac{1}{R\left(\beta_i\right)},$$

where R is a monotonically increasing function of the total benefit $\beta_i := \sum_k B\left(\sigma_{ik}\right)$, in the simulations chosen as $R(\beta_i) := 1 + \nu\beta_i$. As above, we assume the benefit function B to be sigmoidal. Moreover, we assume the cost function to be super-linear and of the general form $C\left(\Sigma_i\right) \propto \left(\Sigma_i\right)^\gamma$.

Below, we show that in the adaptive model property (ii) still holds while property (i) needs to be modified: The total amount of investment differs among agents within a BCC as agents enjoy benefit-dependent cost reduction (Figure 3.9). However, we find that agents of the same degree approach the same investment level. Consequently distinct classes of agents arise which differ both, in investment and in payoff [163].

For deriving the modified coordination properties (i) and (ii) we proceed analogously to the non-adaptive case, i.e., we evaluate the conditions for a solution of the ODE system (3.3) to be stationary and stable.

First the stationarity condition. Defining $\partial_x := \partial/\partial x$, we can rewrite the stationarity condition

$$\frac{d}{dt}e_{ij} = 0 = \partial_{e_{ij}}\left[\sum_k B(\sigma_{ik}) - \frac{C(\Sigma_i)}{R(\beta_i)}\right] \tag{3.23}$$

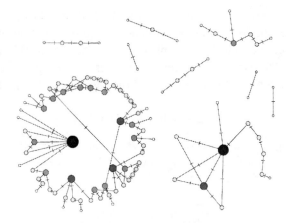

Figure 3.9 Example of communities formed in the adaptive model with 100 players. The communities are dominated by leaders, which are characterized by high degree and high payoff. The size of a node indicates an agent's total investment Σ_i. Within each component, agents with the same degree maintain the same Σ_i. (Parameters: $\rho = 0.7$, $\tau = 0.1$, $\mu = 2.24$, $\nu = 0.588$)

as

$$\partial_{e_{ij}} B(\sigma_{ij}) = \frac{\partial_{e_{ij}} C(\Sigma_i)}{R(\beta_i)} - \frac{C(\Sigma_i)}{(R(\beta_i))^2} \partial_\beta R(\beta_i) \partial_{e_{ij}} \beta$$

$$= \frac{R(\beta_i)}{(R(\beta_i))^2 + \partial_\beta R(\beta_i)} \frac{\partial_{e_{ij}} C(\Sigma_i)}{C(\Sigma_i)}, \tag{3.24}$$

where we used $\partial_{e_{ij}} \beta = \partial_{e_{ij}} B(\sigma_{ij})$. The right hand side of equation (3.24) does only depend on the node parameters Σ_i and β_i, i. e., in a steady state $\partial_{e_{ij}} B(\sigma_{ij})$ is identical for all bilateral links ij of agent i. From $\partial_{e_{ij}} B(\sigma_{ij}) = \partial_{e_{ji}} B(\sigma_{ij})$ it then follows that all bilateral links of j, and, by iteration, that all bilateral links within one BCC are identical with respect to $\partial_{e_{mn}} B(\sigma_{mn})$. Since the benefit function is sigmoidal, a given slope can be found in at most two points along the curve: one above and one below the inflection point (IP) (cf. Figure 3.3). This implies that if a stationary level of investment is observed in one link, then the investment of all other links of the same BCC is restricted to one of two operating points.

In the basic model, stability analysis revealed that the operating point below the IP is unstable and can thus be ruled out (cf. Figure 3.3). Unfortunately, in the extended model, the analysis cannot be performed to the same extend. However, in extensive numerical simulations we have not observed a single equilibrium which contained a link operating below the IP. This strongly indicates that the dynamics of the extended model are governed by similar stability conditions as the dynamics of the basic model, which reproduces property (ii):

$$\sigma_{ij} \equiv \sigma \; \forall \; \text{links } ij \text{ in a BCC} . \tag{3.25}$$

Combining Eqs. (3.24) and (3.25), we can now derive property (i): Let us consider a single BCC. According to Eq. (3.25), the total benefit of an agent i in this BCC is a function of its degree d_i:

$$\beta_i = \sum_k B(\sigma_{ik}) = d_i \cdot B(\sigma) .$$

Inserting this relation in Eq. (3.24), we find that the left hand side is constant, while the first factor on the right hand side only depends on d_i. The second factor on the right hand side is injective, as we assumed $C(\Sigma_i) \propto (\Sigma_i)^\gamma$. It thus follows that nodes of the same degree have to make the same total investment Σ_i, even if they are only connected through a chain of nodes making different investments. However, nodes of different degree d_i can differ in their total investment.

The emergence of distinct classes of nodes, which differ in degree (and therefore in payoff) and total investment is illustrated in Figure 3.9. The figure shows the final configuration of an exemplary model realization with 100 nodes. Nodes of high degree receive high payoffs (coded in the node color) and run high total investments (coded in the node size).

Compared to the basic model, the adaptive model leads to considerably broadened degree distributions (cf. Figure 3.4). We can thus conclude that the additional resources available to high degree agents are at least in part used to establish additional links. This leads to an increased income disparity in the evolving network.

Let us now address the fairness of individual interactions. As in the basic model, also in the adaptive model the investments of two interacting agents into a common collaboration are usually asymmetric. In Figure 3.9, this is apparent in the position of the fairness indicators on the links: The further it is shifted toward one agent i, the lower the fraction e_{ij}/σ_{ij} that he contributes. Even in small network components, the fairness indicators reveal a flow of investments towards regions of high connectivity; as a general rule high-degree nodes contribute less to an interaction than their lower-degree partner.

In both, the basic and the adaptive model, the specific load distribution in an interaction depends on the exact topological configuration of the respective network component. Hence, for comparing the fairness of load distributions in both models, it is necessary to consider components of the same structure.

The simplest degree-heterogeneous structure is a chain of three nodes i, j and k. In such a structure, the two degree-one nodes i and k necessarily concentrate all their investment in the cooperation with the middle node j, while the latter splits its investment in equal parts. The fraction e_{xj}/σ_{xj} that the middle node contributes to each of the two links can be calculated as

$$\frac{e_{xj}}{\sigma_{xj}} = \frac{0.5\Sigma_j}{0.5(\Sigma_i + \Sigma_j + \Sigma_k)}, \quad x = i, k .$$

In the basic model, the total investment of all three nodes are identical. Thus, $e_{xj}/\sigma_{xj} = 1/3$. In the adaptive model, $\Sigma_j > \Sigma_i = \Sigma_k$. Thus, $e_{ij}/\sigma_{ij} > 1/3$, i.e., the load distribution is fairer than in the basic model (cf. fairness indicators on three node chain in Figure 3.9).

Generalizing the reasoning sketched above, we find that for any given topological configuration, the imbalance in the load distribution is milder in the adaptive model than in the basic model. We can thus conclude that the additional resources available to high degree agents are partly reinvested in existing links enhancing the fairness of the respective interactions.

Further confirmation for fairer load distributions in the adaptive model comes from the numerical data: In extensive simulations using a wide range of parameters we have not observed a single unidirectional link. This observation stands in sharp contrast to the observations made in the basic model, where unidirectional links – the most extreme case of unequal load distribution – constitute a considerable fraction of all links in a network.

3.9 Discussion

In this chapter, we have analyzed a model for the formation of complex collaboration networks between self-interested agents. In this model, the evolving network is described by a large system of deterministic differential equations allowing agents to maintain different levels of cooperation with different partners.

We showed analytically that bidirectionally communities are formed, in which every agent makes the same total investment and every collaboration provides the same benefit. In contrast to models for cooperation on discrete networks, the present model thereby exhibits a high degree of coordination which can be interpreted as a precursor of a social norm. We emphasized that coordination is generally achieved although single agents possess insufficient information for computing the total investment made by any other agent and although the level of cooperation that is reached in a community is not fixed rigidly by external constraints.

Despite the high degree of coordination, we observed the appearance of privileged agents, reminiscent of the leaders emerging in [142]. In the model proposed in the present chapter, the privileged agents hold distinguished topological positions of high degree centrality allowing them to extract much higher payoffs than other agents, while making the same cooperative investment. However, we found that in the absence of further mechanism reinforcing differences the assembled topologies were fairer than random graphs.

Although our primary aim was to investigate the formation of social networks, some aspects of the behavior of social agents are reminiscent of results reported in psychology. For instance, our investigation showed that agents can react to the withdrawal of investment by a partner either by mutual withdrawal of resources or by reinforcing the collaboration with increased investment. Our analysis provides a rationale which links the expected response to the withdrawal of resources to an inflection point of an assumed benefit function.

Furthermore, we investigated under which conditions non-reciprocated collaborations appear. Here, our analysis revealed that such unidirectional collaborations can appear in three distinct scenarios, which can be linked to topological properties of the evolving networks. In particular, exploited agents whose investments are not reciprocated invest less than the average amount of resources in their links when occurring in small components, but more than the average amount, when integrated in large components.

If cost reduction for successful agents makes additional resources available to highly-connected agents, these resources are partly invested in existing collaborations, leading to fairer load distributions, but also in establishing new collaborations, leading to broadened degree distributions.

The results reported here can be verified in laboratory experiments, in which humans interact via a computer network. Inspired by our model, Fehl et al. ran an experiment, in which the players could adopt different behavioral options toward different, self-chosen partners [162]. In contrast to our model, however, the behavioral options in the experiment are discrete; players can either cooperate or refuse to cooperate with a given partner. We believe that an experiment with continuous behavioral options may confirm the topological properties of the self-organized networks reported here. Additionally, it may provide insights into the perceived cost and benefit functions that humans attach to social interactions.

Furthermore, results of the proposed model may be verified by comparison with data on collaboration networks between people, firms or nations. This comparison may necessitate modifications of the model to allow for instance for slightly different cost functions for the players. Most of these extensions are straightforward and should not alter the predictions of the model qualitatively. For instance in the case of heterogeneous cost functions, players will make different total investments, but will still approach an operating point in which the slope of their cost function is identical. Further, coordination should persist even if the network of potential collaborations is not fully connected. Finally, but perhaps most importantly, our analytical results do not rely heavily on the assumption that only two agents participate in each collaboration. Most of the results can therefore be straightforwardly extended to the case of multi-agent collaborations.

Our analytical treatment suggests that the central assumption responsible for the emergence of coordination is that the benefit of a collaboration is shared between the collaborating agents, but is independent of their other collaborations, whereas the cost incurred by an agent's investment depends on the sum of all of an agent's investments. Because this assumption seems to hold in a relatively large range of applications we believe that also the emergence of coordination and leaders by the mechanisms described here should be observable in a wide range of systems.

4

Self-organized criticality

The emergent, collective phenomena addressed in the last chapters were dynamical and structural in nature. In this chapter, we consider the self-organization into a functional state.

To approach the topic, recall that a system, which depends on external parameters, may have different phases distinguished by qualitatively different dynamics. Particularly in biology, however, dynamics are intimately related to function. One may thus ask, if a biological system has to operate in a certain phase for optimizing its functionality. Central in this field is the so-called criticality hypothesis [169], according to which critical dynamics close to a phase transition provide particular functional gains.

Early evidence for the criticality hypothesis came from the field of cellular automata [170]. In these systems, the tuning of a control parameter commonly changes the macroscopic behavior from highly ordered to chaotic. The transition between the respective dynamical phases occurs in a narrow parameter regime, sometimes called the edge of chaos, in which the cellular automata display optimal adaptation and information-processing properties [170].

A number of recent papers have yielded a comprehensive picture of the relation between critical dynamics and computational properties [171, 172]. The investigations are based on different model systems and achieve a high level of generality by linking fundamental characteristics of critical states to fundamental aspects of information theory. Information is generally assumed to be coded in the dynamical attractor that is reached in response to an input. Thus, the maximal number of attractors at criticality has been associated with maximal information storage capacity [173, 174], and the diverging correlation length at criticality with maximal information transmission [175]. It has been emphasized that the scale invariance of observables bears the potential to code information over several orders of magnitude [176]. And finally, it has been shown that fluctuations and spontaneous activity can assist the processing of stimulus-evoked activity and thus broaden the range of processable inputs [177].

On the background of the sketched results, it is plausible to assume that our brain may operate at a critical point. Indeed, scale invariance of observables – the hallmark of criticality – was found in different experiments ranging from EEG measurements in humans to direct activity measurements in slices of rat cortex [175, 178–181]. Considering that the brain is subject to ongoing changes throughout development, through aging and damage, it is an interesting question how it preserves the operating parameters prerequisite for criticality. One possible explanation is that the observed criticality is achieved and maintained in a dynamical self-organization process.

The concept of self-organized criticality (SOC) was first proposed by Bak, Tang, and Wiesenfeld. In their seminal work [182], the authors presented a simple cellular automaton, which evolves into a state that features several characteristics of natural critical systems, such as fractal geometry, $1/f$ noise, and scale invariance. The critical state is reached due to the intrinsic dynamics of the system, independently of the value of any model parameter.

In recent years, it has been shown that beside cellular automata, also adaptive networks have the ability to self-organize toward a critical state [19, 155, 156, 183–191]. Indeed, it was found that SOC is not only a generic, but also a robust feature of the latter model class. Today, adaptive SOC is studied with different perspectives: The hope is that understanding the phenomenon in simple models will help to comprehend SOC in biological systems, as well as to implement SOC in technical systems. One step in the latter direction shall be made in the present chapter.

We begin in Section 4.1 with a short overview of the first generation of SOC models, and describe the common mechanism which has been found to underlie them. Then, in Section 4.2, we contrast these models with the later models based on adaptive networks. Condensing the latter to their common core reveals that on an abstract level the genesis of SOC in adaptive networks is linked to the adaptive feedback loop. In Section 4.3, we shortly review four exemplary models showing adaptive SOC. We highlight both, the conceptual similarities in the basic design, as well as the specific differences in their concrete rules. In Section 4.4, we derive a generic recipe for the construction of rules that generate SOC. Finally, in Section 4.5, we use the recipe to design an adaptive network of heterogeneous phase oscillators that self-organizes toward the onset of synchronization.

4.1 SOC models of the first generation

Initiated by Bak et al. [182], a wave of cellular automata models have been studied under the paradigm of SOC (cf. [192–196] and references therein). They all consider non-equilibrium systems, in which constant microscopic driving leads to series of macroscopic events, so-called avalanches, whose frequency and size follow power-law distributions [196].

By way of illustration, consider the sandpile model studied in [182]. In this model, grains of sand are placed onto random sites of a lattice. The so-defined microscopic driving builds up the slope of the pile. When the slope at a site exceeds a certain threshold, it transfers sand to the adjacent sites. This potentially causes cascading reactions, i.e., avalanches, which are found to obey power-law statistics.

According to Dickmann et al. [197], SOC in the studied models can generically be traced back to the systems' approaching a configuration at a continuous absorbing-state phase transition. Below, we describe the mechanism through which this is achieved on an intuitive level without addressing the mathematical subtleties of the argument.

Consider a cellular automaton, whose sites are either active or passive with respect to a local dynamical rule, which we call the avalanche rule. The density of active sites, ρ, is an order parameter of an absorbing-state phase transition. It is zero in states where no avalanche can occur (frozen regime), and non-zero otherwise (active regime). The avalanche rule dynamically changes ρ; Active site are dissipated in avalanches, such that in the active regime the system is driven toward the frozen regime. To achieve SOC, a second dynamical rule is needed, which dominates in the frozen regime and drives the system back toward the active regime. One possible realization of this is a slow driving that increases the density of active sites.

In summary, the genesis of SOC in cellular automata builds on two ingredients: An absorbing phase transition and two contrary dynamical rules effecting

$$\dot{\rho} < 0 \;\; \text{if} \;\; \rho > 0 \;\;\; \wedge \;\;\; \dot{\rho} > 0 \;\; \text{if} \;\; \rho = 0 \,, \tag{4.1}$$

such that the order parameter ρ is tuned to a critical point, where its value changes from zero to finite.

Note that the tuning parameter in Eq. (4.1) is identical to the order parameter. The system has no autonomous parameter by means of which it could be tuned from the frozen to the active phase. Hence, although the critical configuration approached can be associated to a phase transition, it is not possible to draw an order parameter profile.

4.2 SOC in adaptive network models

In 1998, the first-time observation of SOC in an adaptive network model [183] triggered the interest in adaptive networks. Since then, a number of simulation studies have revealed that the ability to self-organize into a critical state is not only a frequent feature of adaptive network models, but also a surprisingly resilient one [19, 155, 156, 184–191]. Thus, in contrast to the SOC models of the first generation, SOC in adaptive network models is very robust to noise, and in many cases even to changes in the modeling setup.

In Section 4.3, we address the robustness of adaptive SOC by means of four exemplary models. Before, however, let us shortly focus on the general definition of the phenomenon, its relation to SOC in cellular automata, and to the adaptive feedback loop.

Prerequisite to adaptive SOC is a network, whose local state dynamics shows different dynamical phases depending on the underlying topological configuration. We speak of SOC if the dynamics *of* the network adjust the topological variables such that the dynamics *on* the network become critical.

The definition of adaptive SOC reveals the differences to SOC in cellular automata. While in the first-generation SOC models the tuning and the order parameter are identical, they are clearly distinguished in adaptive network models: The tuning parameters are the variables of the topological evolution, i.e., the link states. The order parameter, by contrast, characterizes the local evolution on the network, and is thus related to the node states. By tuning the topological variables, the phases to both sides of the approached transition can actually be scanned, and order parameter profiles can be generated.

We can now pinpoint the role of the adaptive feedback loop for the genesis of SOC. Due to the feedback from the topology to the local dynamics, the local dynamics are parametrized by the topological configuration. Due to the feedback from the local states to the topological dynamics, however, the change of the topological configuration is sensitive to order-parameter related information. Thus, the space of topological configurations, i.e., the parameter space of the local dynamics, can be navigated in a phase-sensitive manner to find points which lie on the phase boundary.

In the next section, we illustrate the typical implementation of the adaptive feedback by means of four models showing adaptive SOC. Though related, the models display considerable differences in their specific rules thus providing an example for the robustness of the phenomenon against variations of the actual setup.

4.3 Four examples for adaptive SOC

The most prominent model system for adaptive SOC is the brain. Its activity shows signatures of criticality, although the underlying synaptic connectivity is too complex to possibly be hard coded in the genome [200]. Indications for the decentralized shaping of neural connections based on local, activity-dependent mechanisms have been found in various experimental settings [201–203]. Two of the proposed mechanisms – homeostatic plasticity [204] and spike-time-dependent plasticity [205] – were subsequently studied in AN models. These models revealed that homeostatic plasticity tunes a system toward the onset of self-sustained activity [19, 185–187, 189], and spike time dependent plasticity toward the onset of synchronous activity [190].

Below, we briefly review four models which address the self-organization of neural networks under the action of homeostatic plasticity. We begin our discussion with two models studied by Levina et al. [186], and Meisel respectively [189]. Both models consider an adaptive network, whose links correspond to synapses and whose nodes are modeled as integrate-and-fire neurons: The node state, corresponding to the neuron's membrane voltage, is represented by a continuous variable x_i. The state variable is incremented, if the node perceives an incoming spike. Once it exceeds a threshold, the node itself emits a spike. Meisel additionally includes a leakage term, according to which the state variable decays, if a node does not receive an incoming spike.

When simulated on static topologies, the described dynamical rules give rise to qualitatively different long-term behaviors, in which the spiking activity is either sustained (active phase) or suppressed (frozen phase). The probability of observing persisting macroscopic activity depends on the topological configuration of the underlying network. It increases rapidly, if the average connectivity of the network exceeds a critical value [206].

In the two models under consideration, the topological configuration is dynamically tuned. The tuning results in both cases in critical configurations at the transition between the frozen and the active phase, although it is implemented in different ways: Levina et al. consider directed links with continuous weights. If a node spikes, the weights of all outgoing links are decreased; if it does not spike, the link weights slowly regain their default strength [186]. By contrast, Meisel considers directed links with discrete weights $\pm g_c$. If a node spikes, it looses all outgoing links; if not, it gets a new link from a randomly chosen node [189].

The two models of Levina et al. and Meisel, respectively, provide a first example for the robustness of adaptive SOC against variations in the modeling setups: Their differences, manifest in the rescaling of the weights of fixed connections compared to the reformation of connections with binary values, does not affect the genesis of SOC.

The two models we want to address next are not as closely patterned on neuronal systems as the afore-discussed, but can nevertheless be considered to map the action of homeostatic plasticity in the brain. In the adaptive networks considered by Bornholdt and Rohlf [19], and Bornholdt and Röhl [185], the node evolution is determined by boolean threshold dynamics. Thus, in [19], nodes and links can adopt binary states $x_i = \pm 1$, $A_{ij} \pm 1$ respectively. In every timestep t, every node i sums over the inputs $A_{ji}x_j(t)$ it receives from its neighbors j. If the result is smaller (larger) than a threshold h, it adopts state $x_i = +1$ ($x_i = -1$) in the next timestep.

Just as the integrate-and-fire neuron model, the boolean threshold model has a phase transition from a frozen to an active phase, depending on the underlying topological configuration. And just as the neuron model, the threshold model can self-organize toward a critical configuration if run in an adaptive scenario. In [19], Bornholdt and Rohlf introduce adaptivity by prescribing the repeated application of the following topological update rule: The local dynamics of the system are simulated on a static topology until a dynamical attractor such as a limit cycle has been reached. Then, a randomly chosen node changes its topological environment in one of two possible ways depending on a local activity measure: If the node has changed its state at least once over the course of the attractor, it deletes an incoming link, if not, it gets a new incoming link from a random node.

The model studied by Bornhold and Röhl [185] is a variation of [19]. Main modifications concern the introduction of continuous link states $A_{ij} \in [-1, 1]$, and of stochasticity in the node dynamics. Finally, the topological update rule considers not only one randomly chosen node i, but a pair of randomly chosen nodes i and j. The absolute correlation $C_{ij} = \left| \frac{1}{\tau} \sum_{t=t_0}^{t_0+\tau} x_i(t)x_j(t) \right|$ of i and j, averaged over a fixed, long period τ, serves as a criterion for i to opt between the two modes of topological adaptation: If C_{ij} is larger than a given threshold, i receives a new link from node j, if it is smaller, the link between i and j is deleted.

Here, as above, the differences in the setup do not affect the genesis of SOC: Both models, as well as a number of other variations described in [185], show robust self-organization toward a critical state.

In all four models, the topological update rule captures two opposing processes, e.g., link building and link deletion, which are applied depending on some local measure that relates to the node state dynamics. Below, we expand this pattern for formulating a generic recipe for the construction of topological update rules that give rise to SOC. While the recipe highlights on the one side common design principles, it classifies on the other hand the possibilities to modify the concrete setup without affecting SOC.

4.4 Engineering adaptive SOC

Below, we consider the following abstract scenario: Given a local update rule that gives rise to distinct macroscopic phases when run on different static topologies, can we construct a topological adaptation rule such that the resulting adaptive network exhibits SOC? We start our considerations by specifying three conditions, which the adaption rule shall meet beyond generating SOC. These conditions are imposed to account for the above given definition of self-organization.

Firstly, the adaptation rule shall act on *local topological variables*. That is, the adaptation pertains either to single links, or to a set of links connected to a focal node. For the moment, we only consider the case of continuous variables. Possible choices for a continuous and local topological variable include the weight of a single link, or the total weight of all links of a focal node. Below, the number of topological variables in the network will be denoted by d, and the variables themselves by T_i, $i = 1, \ldots, d$.

Secondly, the adaptation of the topological variables shall occur solely on the basis of *local information*. This concerns the implementation of the feedback from the node states to the topological dynamics. Local information about the dynamics on the network is information that a focal node can gather by monitoring its own state and / or the influences it experiences from its direct neighbors. Below, such information will be denoted by O_i if accessible to a node i.

Thirdly, all topological variables shall be subject to the *same* adaptation rule. This is done to account for the idea that the constituents of a self-organizing system are structurally identical. By way of illustration, consider the model of Bornholdt and Rohlf described above [19]. In this model, the topological update rule is the same for all nodes, while the topological variable, to which it is applied, and the local information, on which it depends, are node specific.

In Section 4.3, we have seen four examples for topological adaptation rules that meet the imposed conditions. They all follow a pattern according to which a local topological variable T_i is changed in one of two possible ways depending on some local measure O_i. Below, we first ask which underlying

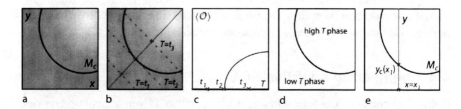

Figure 4.1 Steering the space of topological configurations: Schematic illustration using an example with $d = 2$. (a) Shown is the two-dimensional space of topological configurations spanned by the variables $T_1 = x$ and $T_2 = y$. The order parameter profile, coded in the coloring of the plane, divides the space of topological configurations in two different phases. The kink in the profile defines the phase boundary M_c (depicted as black line), which is of codimension one. (b) In numerical studies, random graphs with different configurations (x, y) and (x', y') are collapsed onto one data point t_i, if they satisfy $T(x, y) = T(x', y') = t_i$ for some function $T : \mathbb{R}^2 \rightarrow \mathbb{R}$. Geometrically, this means that a hypersurface characterized by $T = $ const., is projected onto one point. The figure shows three hypersurfaces (dashed grey lines), and the projection manifold T (solid grey line). The statistical analysis of the ensemble data associates to every value of t_i the average value of the order parameter $\langle \mathcal{O} \rangle$. (c) Plotted is the resulting one-dimensional order-parameter profile. The minimal t, for which the phase transition is observed (here t_2), is usually referred to as critical value t_c, at which the transition occurs. Note, however, that the terminology is misleading, as $T(x, y) = t_c$ cannot be viewed as a universal characteristic of all critical configurations. (d) The statistical projection of (c) can be used to distinguish a high-T, and a low-T phase. (e) In a sequential update, only one degree of freedom is updated in every given adaption step, and all others are kept constant. Sketched is a situation, in which y is tuned, while x is fixed at $x = x_1$. By tuning y, the configuration (x, y) may be shifted along the red line. The intersection point $y_c(x_1)$ of this line and M_c is approached if y is increased (decreased) when the system is in the high (low) T phase.

principles make this pattern tune a system toward a critical configuration. Then, we analyze which properties of the O_i and the T_i are necessary for the pattern to work. This finally allows us to expand the pattern to a generic recipe for the construction of adaptation rules generating SOC.

How and under which conditions does the alternative application of two opposing processes tune a system toward a critical configuration? For the sake of simplicity, we approach the question in a discrete time scenario. As a starting point, we use the generic properties of the prerequisite phase transition of the local state dynamics. We may generically assume that it can be observed in ensembles of random graphs, which differ in the macroscopic parameter

$$T = \frac{1}{d} \sum_{i=1}^{d} T_i \qquad (4.2)$$

that subsumes all adaptable degrees of freedom (cf. Figures 4.1a-c). In this case, the two phases can uniquely be described as low-T, high-T phase respectively (cf. Figure 4.1d). Depending on the definition of the T_i, T may for instance denote the average coupling strength in the network.

The codimension of a transition that can directly be observed, and thus the minimal number of degrees of freedom that need to be adapted for arriving at the phase boundary M_c is one. Hence, it is convenient to determine that in every step of the topological evolution only one randomly chosen variable T_i is updated, while all others (although in principle dynamic) are kept constant. In the literature, this is often called a *sequential update*.

The sequential update decomposes the process, which navigates the d-dimensional space of topological configurations, in a series of processes that are one-dimensional; in each evolutionary step, the randomly chosen variable T_i can only be increased or decreased. Depending on the values of the fixed variables T_j, $j \neq i$, the updated variable T_i can have a critical value $(T_i)_c$, which geometrically corresponds to an intersection point of the $(d-1)$-parameter family T_i and M_c (cf. Figure 4.1e). If $(T_i)_c$ exists, it can be approached, by increasing (decreasing) T_i, if the system is in the low (high) T phase. If $(T_i)_c$ does not exist, the application of same rule increases the probability that in the next evolutionary step, when a different variable T_k is updated, it finds parameters T_j, $j \neq k$, such that $(T_k)_c$ exists.

Let us shortly summarize. The above reasoning illustrates how the d-dimensional space of topological configurations can be steered by means of a sequential update rule that prescribes the phase-dependent application of two

One example for an order parameter of the form Eq. (4.4) is the order parameter $\mathcal{O}$ that Bornholdt and Rohlf use to characterize the absorbing phase transition of the threshold model [19]. In this paper, $\mathcal{O}$ is defined as the fraction of nodes that has not changed state over the course of an attractor. This is equivalent to the population average over a local measure O_i, which is 1 if a node i has not changed state, and 0 if it has.

Another example is the order parameter that Meisel and Gross use to characterize the synchronization transition of the integrate-and-fire neuron model [190]. In this paper, the definition of $\mathcal{O}$ is based on a local measure $C_{ij}(\tau)$, which denotes the correlation between the state of node i and node j over a period τ. The order parameter $\mathcal{O}$ is the mean value of $C_{ij}(\tau)$ average over all pairs i, j.

One example for an order parameter, which cannot be understood as population average over a local measure, is the order parameter commonly used to characterize the synchronization transition of the Kuramoto model. It is defined as the norm R of the vector describing the centroid of the population of phase oscillators, and can be written as:

$$ R = \left| \frac{1}{N} \sum_{j=1}^{N} e^{ix_j} e^{-i\psi} \right| = \left| \frac{1}{N} \sum_{j=1}^{N} e^{i \sum_{k \neq j}(x_j - x_k)} \right| , \qquad (4.3) $$

where x_j is the phase of oscillator j, and $\psi = 1/N \sum_{j=1}^{N} \omega_j$ is the mean of the intrinsic frequencies. Inserting the defining relation of ψ, the coordinates of the centroid can be rewritten as population average over a local measure accessible to every node j. However, the norm of the vector cannot be rewritten as average over locally computable quantities.

Box 4 Global order parameters are often population averages over local measures.

opposing processes. It further reveals that the variables T_i and the local measure O_i employed by such a rule have to meet the following requirements: The topological variables T_i have to be chosen such that the phases can be uniquely be described as high-T phase, low-T phase respectively. The local measure O_i has to allow for the detection of the global phase such that it can be used as local criterion for the phase-specific application of the opposing processes. As the condition on the T_i is rather mild, we below concentrate on

To demonstrate the application of the homogeneity definition Eq. (4.5), we again use the example of the absorbing phase transition of the threshold model [19]. In particular, the example shall illustrate that homogeneity is not a property that pertains to a phase as such, but a property that depends on both, the phase and the exact definition of the O_i. We begin with defining the local measure O_i as the number of times a node i has changed state over a period τ; $\mathcal{O} = \sum_i O_i$ is an order parameter of the phase transition and is zero in the frozen and non-zero in the active phase. The definition of the frozen phase requires that in this phase all nodes i measure $O_i = 0 = \mathcal{O}$. Hence, the frozen phase is homogeneous. The definition of the active phase requires that in this phase at least some nodes measure $O_i \neq 0$. We show below that the number of nodes measuring $O_i \neq 0$, and therewith the homogeneity or inhomogeneity of the active phase, depends on the particular value of τ.

In the active phase, the local threshold dynamics are ergodic [207, 208]. This implies that in this phase a randomly chosen node changes state with probability one if monitored over an infinite period of time, and with a probability close to one, if monitored over an finite but long period τ. Thus, if τ is chosen sufficiently large, all nodes i measure $O_i \neq 0$ in the active phase. In this case the active phase is homogeneous.

On the other hand, the probability of finding one or more nodes i measuring $O_i = 0$ increases with decreasing τ. Thus, if τ is chosen sufficiently small, the active phase is inhomogeneous.

Box 5 The homogeneity of a phase depends on the exact definition of O_i.

the task of finding local measures O_i, by means of which the global phase can be detected.

Consider a local measure O_i, whose global average is an order parameter $\mathcal{O}$ of the phase transition. Finding such a measure O_i is typically easy, as standard order parameters can often be rewritten as

$$\mathcal{O} \propto \sum_i O_i \, . \tag{4.4}$$

For examples we refer to Box 4. Next, we have to ask under which conditions the local O_i, as the global $\mathcal{O}$, contain information about the global phase. For addressing this question, it will prove convenient to introduce the following conventions: If for a given phase and a given order parameter of the form (4.4) holds

$$\mathcal{O} = 0 \;\Rightarrow\; O_i = 0 \;\forall i \;, \tag{4.5a}$$

$$\mathcal{O} \neq 0 \;\Rightarrow\; O_i \neq 0 \;\forall i \;\text{respectively}\;, \tag{4.5b}$$

we call the phase *homogeneous* in the O_i. Analogously, if the applicable relation is violated, the phase is called *inhomogeneous* (cf. Box 5).

We can now distinguish three different cases: Case (a), in which both phases of a system are homogeneous, case (b), in which both phases of a system are inhomogeneous, and case (c), in which one phase is homogeneous, while the other is inhomogeneous.

In case (a), the local observation O_i is unconditionally equivalent to the global observation $\mathcal{O}$. Hence, the complete information about the global phase is locally accessible to all nodes and in both phases.

In case (b), there is no local observation O_i that could reliably be linked to a global observation $\mathcal{O}$. Thus, the information about the global phase cannot locally be accessed via O_i.

Finally, in case (c), only one of the two possible local observations O_i can reliably be linked to a global observation $\mathcal{O}$: The local observation O_{hom} that is made by all nodes in the homogeneous phase is measured also by some nodes in the inhomogeneous phase. Thus, it does not convey information about the global phase. By contrast, the local observation $O_i \neq O_{\text{hom}}$ conveys that the system is in the inhomogeneous phase. Hence, the global phase can locally be detected via O_i, albeit the detection is restricted to the inhomogeneous phase, and, within this phase, to nodes measuring $O_i \neq O_{\text{hom}}$. It is worth noticing that the fraction f of nodes that measure $O_i \neq O_{\text{hom}}$ is an order parameter of the transition.

We can now summarize the conditions under which the alternative application of two opposing processes tunes a system toward a critical configuration: Our analysis revealed that it is convenient to implement the update rule as a sequential rule, and that it is necessary to make the application of the opposing processes phase-specific. We found that the topological variables T_i should allow for the unique definition of a low-T and a high-T phase, and that the process that prevails in the high-T (low-T) phase should decrease (increase) T_i. We found further that a local phase-detecting criterion O_i should have a global mean that is an order parameter of the transition, and lastly that it is necessary to define O_i such that at least one of the phases is homogeneous. Below, we finally adopt the engineering viewpoint and translate the conditions on the T_i, O_i, and the opposing

> *Given a system, whose local state dynamics have two different dynamical phases depending on the underlying topological configuration,*
>
> 1. *Find a local measure O_i, whose global mean $\mathcal{O} \propto \sum_i O_i$ is an order parameter of the transition.*
> 2. *Check for each phase whether it is homogeneous or inhomogeneous in O_i. If at least one phase is homogeneous, proceed.*
> 3. *Define local topological variables T_i. If the local measure O_i pertains to a node i (a link i), T_i should capture the weights of the links of a node i (the weight of a link i).*
> 4. *Sample random graphs with different $T = \sum_i T_i$ and measure $\mathcal{O}$ to determine high-T / low-T phase.*
> 5. *Relate local measurements $O_i = 0$ and $O_i \neq 0$ to high-T / low-T phase.*
> 6. *Stipulate that in every step of the topological evolution, one T_i is chosen at random. It is decreased by $\delta_\downarrow$ (increased by $\delta_\uparrow$) if O_i indicates that the system is in the high-T (low-T) phase.*
> 7. *If one of the phases is inhomogeneous, choose the rate of change of the process applied for $O_i \neq O_{hom}$ larger than the rate of the process applied for $O_i = O_{hom}$.*

Box 6 Recipe for a topological update rule generating adaptive SOC

processes into a recipe for their construction.

To define a local measure O_i that is a suitable criterion for the phase-specific application of the two opposing processes, we can proceed as follows: First, we have to find a measure O_i, whose global mean $\mathcal{O} \propto \sum_i O_i$ is an order parameter of the transition. Second, we have to check for each phase whether it is homogeneous or inhomogeneous in O_i. Third, if either one, or both phases are homogeneous in O_i, the measure can be applied as a switching criterion between the opposing processes: In case both phases are homogeneous, the two possible measurements $O_i = 0$ and $O_i \neq 0$ can uniquely be related to the high-T phase, low-T phase respectively. On this basis, we can then specify whether a given topological degree of freedom T_i shall be increased or decreased depending on the value of O_i.

In case one phase is inhomogeneous, only the measurement $O_i \neq O_{hom}$ is a reliable indicator of the global phase and can uniquely be related to

either the high-T or the low-T phase. The bounded information content of
the measurement $O_i = O_{hom}$ can be compensated if the opposing topological
processes are implemented with different rates of change. Thereby, the process that is applied for $O_i \neq O_{hom}$ should have a larger rate of change than
the process that is applied for $O_i = O_{hom}$.

In Box 6, we summarize the above considerations in a recipe for the
construction of a topological update rule generating SOC. To evaluate the
effect of such a rule, we below study one scenario quantitatively.

Consider a system, whose local state dynamics have two different phases
depending on the topological configuration. Consider further a local measure
O_i, with respect to which one the phases is homogeneous, the other inhomogeneous, and a set of local variables T_i such that the homogeneous phase
is the low-T phase. Let us next define $\delta_\uparrow$ and $\delta_\downarrow$ as rates of change of the
processes applied for $O_i = O_{hom}$, $O_i \neq O_{hom}$ respectively. If we moreover
adopt a continuous time picture, the balance equation for each topological
variable can be written as differential equation

$$\frac{d}{dt}T_i = \begin{cases} \frac{1}{d} \cdot \delta_\uparrow & \text{in the homogeneous phase,} \\ \frac{1}{d} \cdot [1 - f(\mathbf{T})] \cdot \delta_\uparrow - \frac{1}{d} \cdot f(\mathbf{T}) \cdot \delta_\downarrow & \text{in the inhomogeneous phase.} \end{cases}$$

(4.6)

Here, $\mathbf{T} = (T_1, \ldots, T_d)^t$ denotes the d-dimensional vector describing the
topological configuration, while $f(\mathbf{T})$ denotes the configuration-dependent
fraction of nodes measuring $O_i \neq O_{hom}$. The term $\frac{1}{d}$ corresponds to the
probability that the variable T_i is chosen for an update. With probability
$f(\mathbf{T})$, the chosen node measures $O_i \neq O_{hom}$ and thus decreases T_i by $\delta_\downarrow$,
with probability $1 - f(\mathbf{T})$, it measures $O_i = O_{hom}$ and thus increases T_i by
$\delta_\uparrow$.

We can immediately read off the implicit function defining the stationary
points $\mathbf{T}^*$ of Eq. (4.6)

$$f(\mathbf{T}^*) = \frac{\delta_\uparrow}{(\delta_\uparrow + \delta_\downarrow)}.$$

(4.7)

The left hand side denotes the value of the order parameter f in a stationary
topological configuration. The right hand side may in the continuous time
picture be considered as a measure for the timescale separation between the
topological processes. As the right hand side of Eq. (4.7) is non-zero for
finite $\delta_\uparrow$, $\delta_\downarrow$, all stationary configurations lie in the inhomogeneous phase.
Assuming that, in this phase, f increases with increasing distance to the
phase boundary, we can infer that the steeper f and the larger the separation between the two timescales of the topological evolution, the closer the

fixpoints $\mathbf{T}^*$ lie to the phase boundary. Further, it follows from the same assumption that all stationary configurations $\mathbf{T}^*$ are stable, and, thus, that the system converges to one of them.

Equation (4.7) reveals that the influence of the parameters $\delta_\uparrow$, and $\delta_\downarrow$ on the assembled topological configuration is the smaller, the sharper the transition of the order parameter f. In particular, the solution $\mathbf{T}^*$ of the stationarity condition (4.7) becomes independent of the right hand side if the fraction of nodes measuring $O_i \neq O_{hom}$ jumps sharply from 0 to 1 at the transition. This corresponds to the limit, in which the inhomogeneous phase becomes homogeneous, i.e., to the limit, in which case (c) becomes case (a). As shown above, the transition from case (c) to case (a) is related to the long-term accumulation of the local measure O_i if the dynamics in the inhomogeneous phase are ergodic [19, 185]. We may thus say that in this case the timescale separation between the local and the topological dynamics substitutes the timescale separation between the two topological processes.

The two different modes of implementing the timescale separation may be illustrated using the examples given in Section 4.3: In the models of Levina et al., and Meisel, the accumulation time τ of the local activity measure is comparably short [186, 189] such that the active phase is inhomogeneous. Indeed, in accordance with Eq. (4.7), the rate of change in the inhomogeneous, active phase exceeds in both models the rate of change in the homogeneous, frozen phase.

In the models of Bornholdt and Rohlf, Bornholdt and Roehl respectively, the accumulation time τ of the local activity measure is long: In [185], it exceeds the timescale of the local threshold dynamics by a factor of 100. In [19], it is given by the variable length of an attractor. Being one in the frozen phase, the attractor length diverges in the active phase [209], such that in [19], too, the topological dynamics in the active phase are much slower than the local dynamics. Thus, in [19, 185] both phases are homogeneous; the opposing topological processes have the same rate of change: Links are deleted and created at the rate of one link per update.

Note that in the strict sense the topological configurations assembled by our rule are not critical: Due to Eq. (4.7), the fixpoints $\mathbf{T}^*$ lie close to the phase boundary, but not on it. Moreover, due to the statistical nature of Eq. (4.6), a given realization of the process might not converge to a fixpoint $\mathbf{T}^*$, but fluctuate around it. As argued by Bonachela and Muñoz, the difference between configurations that are critical in the strict sense and so-called pseudo-critical configurations is conceptually considerable [199]. However, they also show that it dislimns if primarily the generated phenomenology in finite systems is considered. As we are interested in the latter, we may stick with our terminology. Thus, in our context a system is termed critical if close, but not necessarily arbitrarily close configurations give rise to qualitatively different dynamics.

If we adopt this more qualitative picture, the above argumentation can readily be transferred to the case of discrete topological variables. In this case, the parameter space is not continuous and the concept of the phase boundary looses validity. However, the closeness of configurations can still be defined, for example as minimal number of changes necessary to arrive from one to the other. Hence, systems with discrete topological variables, too, can self-organize toward configurations that are critical in the sense of the definition given above. This applies in particular in the limit of large system sizes.

Let us emphasize that – in contrast to the underlying mechanism of the first generation of SOC models, which we discussed in Section 4.1 – the construction principles determined in this section are not bound to a certain type of phase transition. For demonstration, we use them in Section 4.5 to construct a model that self-organizes toward a synchronization transition.

4.5 Engineering SOC in a model system

In this section, we use the recipe derived in Section 4.4 to construct an adaptive network model that shows SOC. Starting point for our construction is a network, whose local state dynamics are determined by the Kuramoto model. Our adaptive model shall self-organize toward the onset of synchronization.

Let us briefly recall the conventions used. We consider a network of N oscillators i, whose local state evolution is given by

$$\dot{x}_i = \omega_i + \sum_{j \neq i} A_{ij} \sin(x_j - x_i) , \quad \forall i \in 1 \ldots N . \tag{4.8}$$

As above, x_i and ω_i denote the phase and the intrinsic frequency of oscillator i, while $A_{ij} = A_{ji}$ denotes the weight of the link between two oscillators i and j. If $A_{ij} \neq 0$, the oscillators are said to be coupled; if further $\dot{x}_i(t) = \dot{x}_j(t)$ for all t, they are said to be phase locked.

The local, pairwise criterion for phase locking reveals that the Kuramoto model has not only two, but three different phases: The completely disordered phase, where no two oscillators are phase locked, the partially synchronized phase where some oscillators are phase locked, and the complete synchronized phase, where all oscillators are phase locked [210].

The definition of the global phases through the local phase-locking criterion provides an ideal starting point for the construction of our rule. According to Box 6, we first have to find a local measure O_i, whose global mean is an order parameter. This is obviously fulfilled by the measure

$$C_{ij} = \frac{1}{\tau} \int_{t=t_0}^{t_0+\tau} \left[\dot{x}_i(t) - \dot{x}_j(t) \right] dt ,$$ (4.9)

which is zero if two oscillators i and j are phase locked, and non-zero if they are not.

In step 2, we have to compare the local phase-locking criterion C_{ij} with the definition of the global phase. The comparison reveals that the completely disordered and the completely synchronized phase are homogeneous in C_{ij}, while the partially synchronized phase is inhomogeneous:

$$C_{ij} \begin{cases} \neq 0 \text{ for all } i, j & \text{in the disordered phase,} \\ = 0 \text{ for some } i, j & \text{in the partially synchronized phase,} \\ = 0 \text{ for all } i, j & \text{in the completely synchronized phase.} \end{cases}$$ (4.10)

Both phase transitions, the lower transition between the disordered and partially synchronized phase, and the upper transition between the partially and the completely synchronized phase, separate one homogeneous and one inhomogeneous phase. Thus, C_{ij} can be employed as a local, phase-detecting criterion O_i for a topological update rule that tunes the system toward either of them.

In step 3, we have to define the topological variables that shall be adapted by our rule. As the chosen local measure $O_i = C_{jk}$ pertains to a single link, it is convenient to let the adaptation processes, too, act on a single link. Thus, we define the variables T_i as link weights A_{jk}.

If the average coupling strength $T = 1/d \sum_i T_i$ of a network is incrementally increased, the three phases are observed in the order of Eq. (4.10) [210]. Hence, the disordered phase is the low-T phase, the partially synchronized phase the intermediate-T phase, and the completely synchronized phase the high-T phase of our system.

Next, we have to prescribe which of the opposing processes – increasing or decreasing A_{ij} – shall be applied for which values of C_{ij}. For this purpose consider that irrespective of the regarded phase transition, the local measure $C_{ij} = 0$ is characteristic for the phase with higher T. Thus, we stipulate that for $C_{ij} = 0$ the link weight is decreased by an amount $\delta_\downarrow$, while for $C_{ij} \neq 0$ it is increased by an amount $\delta_\uparrow$.

Finally, we have to specify the rates $\delta_\downarrow$, $\delta_\uparrow$ thereby accounting for the inhomogeneity of the partially synchronized phase. According to step 7 of the recipe, the rate of change of the process applied for $C_{ij} \neq C_{\text{hom}}$ has to be larger than the rate of change of the process applied for $C_{ij} = C_{\text{hom}}$. In case of the upper transition $C_{\text{hom}} = 0$, as the homogeneous of the two abutting phases is the completely synchronized one (cf. Eq. (4.10)). For tuning toward this transition, we thus have to choose $\delta_\downarrow \ll \delta_\uparrow$. In case of the lower transition $C_{\text{hom}} \neq 0$, as the homogeneous of the two abutting phases is the disordered one. For tuning toward this transition, we have to choose $\delta_\uparrow \ll \delta_\downarrow$.

In summary, our topological update rule stipulates that in every evolutionary step one link ij is chosen at random, and the measure C_{ij} is evaluated over a period τ. If $C_{ij} = 0$, A_{ij} is decreased by an amount $\delta_\downarrow$; if $C_{ij} \neq 0$, it is increased by an amount $\delta_\uparrow$. For approaching the lower transition, we choose $\delta_\uparrow \ll \delta_\downarrow$; by contrast, for approaching the upper transition, we choose $\delta_\downarrow \ll \delta_\uparrow$.

To evaluate the defined topological update rule, we study exemplary realizations of the adaptive network model in numerical simulations. In these, we substitute the local measure C_{ij} defined by Eq. (4.9) with a different measure, which can be tracked with less computational effort and has further the advantage of yielding a discrete phase-detecting criterion. Thus, we define C_{ij} as the number of changes of sign that the coupling term $\sin(x_i - x_j)$ displays during a tracked period τ. We verified that if τ is chosen sufficiently large, both described criteria coincide qualitatively.

Our simulations are run on networks of $N = 50$ oscillators, whose intrinsic frequencies are drawn from a Gaussian distribution with mean zero and variance $\sigma^2 = 0.01$. We assume that the network of potential interactions is Erdös–Renyi random graph [211] with mean degree $k = 10$. All links are initially assigned a weight A_{ini}.

Between any two topological updates, we integrate the system of equations (4.8) on a static topology for $\tau = 500$ time units using a Taylor series method [212]. We then choose one random link ij, and evaluate the local measure C_{ij} over the last 0.7τ steps of the integration thus discarding transients. For organizing toward the lower transition, we apply the following rule: If $C_{ij} = 0$, the link weight A_{ij} is decreased by an amount $\delta_\downarrow = 0.01$, but always kept nonnegative; Otherwise it is increased by $\delta_\uparrow = 0.001$. For organizing toward the upper transition, we apply the same rule but with exchanged rates $\delta_\downarrow = 0.001$, $\delta_\uparrow = 0.01$.

Figure 4.2 confirms that the constructed topological update rules tune the system toward configurations, where the order parameter f changes qual-

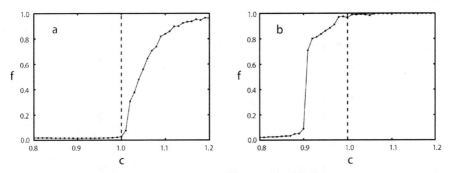

Figure 4.2 One-dimensional order parameter scans around evolved network configurations. Plotted is the average fraction of phase locked links f in topological configurations $\mathbf{T}$ related to the evolved configuration $\mathbf{T}_{\text{evolved}}$ via the scanning parameter c: $\mathbf{T} = c \cdot \mathbf{T}_{\text{evolved}}$. Every data point is averaged over 300 integration runs with different initial phases x_i. In each of these runs, the phase locking criterion C_{ij} was measured over 1000 integration steps. The evolved configuration, marked by the dashed line at $c = 1$, lies close to the lower (upper) transition in case (a), (b) respectively. It results in both cases from 10^4 iterations with the following parameters: $(A_{\text{ini}}, \delta_{\downarrow}, \delta_{\uparrow}) = (0.005, 0.01, 0.001)$ for (a), and $(0.01, 0.001, 0.01)$ for (b).

itatively. The figure shows one-dimensional order parameter scans around two self-organized network configurations. Shown is the average fraction of phase locked links f in topological configurations $\mathbf{T}$ related to the evolved configuration $\mathbf{T}_{\text{evolved}}$ via the scanning parameter c: $\mathbf{T} = c \cdot \mathbf{T}_{\text{evolved}}$. As predicted, the system approaches the lower phase transition if $\delta_{\downarrow} \gg \delta_{\uparrow}$, and the upper phase transition if $\delta_{\uparrow} \gg \delta_{\downarrow}$. Note that in accordance with Eq. (4.7), the evolved network configurations ($c = 1$) lie in both cases slightly in the inhomogeneous phase.

Figure 4.3 shows that the change of the order parameter around the evolved configuration is the steeper, the longer the topology is evolved. This can be understood as follows: Consider a configuration $\mathbf{T}_{\text{evolved}}$ at the boundary of the disordered phase. The change of f around $\mathbf{T}_{\text{evolved}}$ is a measure for the number of links ij, for which the phase-locking criterion C_{ij} changes from $C_{ij} \neq 0$ to $C_{ij} = 0$, if all $A_{ij} \rightarrow (1 + \epsilon)A_{ij}$. As by assumption $\mathbf{T}_{\text{evolved}}$ lies at the phase boundary, the number is at least one. However, there may be points at the boundary, where it is much higher.

Let us now come back to Figure 4.3. It reveals that the phase boundary is reached after 3000–4000 iterations of the topological evolution. The topological evolution toward the phase boundary, however, is followed by a topological evolution along the phase boundary, in the course of which more and more link weights are tuned to critical values. This is a direct

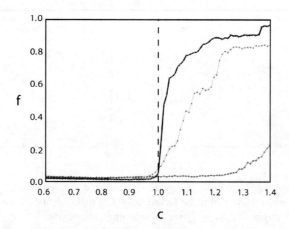

Figure 4.3 Order parameter scans around network configurations, which have been evolved in 3000, 4000, 5000 iterations of topological evolution (grey, light grey, black). The longer the evolution, the steeper the increase of f around the evolved configuration ($c = 1$). Every data point is averaged over 100 integration runs with different initial phases x_i. The evolved networks were obtained with parameters $\left(A_{\text{ini}}, \delta_\downarrow, \delta_\uparrow\right) = (0.0, 0.01, 0.001)$.

consequence of the sequential update rule, according to which topological variables are chosen in an ongoing manner and gradually tuned. Thus, it leads to a configuration which is not only locally distinguished by the proximity to the transition of the order parameter, but also globally by the maximal steepness thereof.

4.6 Discussion

In this chapter, we have analyzed a pattern commonly found in the adaptation rules of heuristic SOC models. By determining the functional principles behind this pattern, we found a generic mechanism that generates SOC. Comparing the specific models with the mechanism allows to analyze the former on an abstract level and to relate the details of their setup to particular functions within the self-organization process. Moreover, the abstract mechanism explains the robustness of SOC against variations in the modeling setup. Thus, it may be considered as set of requirements on the two central entities, T_i and O_i. Within the bounds of these requirements, however, the entities can freely be chosen without affecting the genesis of SOC.

We have shown that the recipe allows to construct systems that self-organize toward criticality. We believe that this engineering viewpoint on

SOC will become particularly important in the future. Thus, for example, as part of the development of nano computing elements the targeted placement of conducting paths gets increasingly demanding. A new and promising approach to this problem is to abstain from targeted placement, put conducting paths at random and implement a neuro-mimetic self-organization process for achieving a functionally connected state. As a positive side effect, such a state can be assumed to feature the favorable information processing properties that we described above.

Engineering solutions that rely on SOC can benefit from the decentralized optimization of computational, but also of structural properties: Thus, as shown in the example of the Kuramoto model, the self-organization process can tune the system toward a state, which is able to support a predefined task, while reducing the required connectivity.

In all models discussed here, the topological update rules tune the total coupling strength and/or the connectivity of the network. However, the analysis of Section 4.4 reveals that SOC can equally be achieved by means of a rule, which acts on other topological structures, for example on the number of triangles. In fact, for the self-organization toward a phase boundary of codimension one, the choice of the topological tuning parameters T_i is quasi unrestricted. Note however that different choices may result in topologically different, albeit dynamically equivalent configurations.

The above reasoning suggests that it is possible to implement systems, that simultaneously self-organize toward more than one phase transition: Consider a dynamical system which features $m < d$ phase transitions, with non-parallel phase boundaries of codimension one. Consider moreover m topological rules, each of which meets the criteria derived in Section 4.4 and depends on a local measure sensitive to one of the considered phase transitions. If the topological update rules steer the parameter space in a non-parallel way, i.e., if they act on different topological structures T_i, it is assured that the individual self-organization processes do not conflict with each other. Hence, their simultaneous application should drive the system to a configuration that lies on the intersection of the m phase boundaries and which is thus critical with respect to all m phase transitions.

5

Conclusions and future research

In this work, we proposed new analytical approaches to adaptive self-organization. These approaches are applicable to continuous networks and allow to address the mechanisms behind emergent phenomena that are prominently discussed in the context of biological systems.

In Chapter 2, we studied the relation between structure and synchronized behavior in a system of coupled oscillators. We introduced a graphical notation, which allows writing the minors of a hermitian Jacobian matrix in a concise way. Imposing a zero-row-sum condition, we reformulated the algebraic stability conditions from Jacobi's signature criterion as topological restrictions on a simple graph G. The topological stability criteria pertain to structures on all scales and provide a rapid test for whether a interaction topology can support stable steady states. In cases where it is violated it allows to determine those interactions that presumably cause the instability.

In Chapter 3, we examined the spontaneous diversification of an initially homogeneous population of interacting agents. Our analysis of the continuous, directed, multi-agent snowdrift game, revealed that the symmetries of the local dynamical rules scale up and are imprinted in non-obvious symmetries in the evolving global structure. These global symmetries imply a high degree of social coordination. However, at the same time they cause the emergence of privileged topological positions, thus diversifying the population into different classes.

In Chapter 4, we addressed the ability of adaptive networks to self-organize toward dynamically critical states. We identified a common pattern in the setup of exemplary SOC models, and laid out how and under which conditions it generates SOC. By expanding the pattern to an abstract mechanism, we were able to explain the robustness of SOC against variations in the modeling setup. Moreover, we were able to formulate a generic recipe for the construction of local rules that give rise to self-organized critical behavior. We demonstrated the applicability of this recipe by engineering an adaptive network of phase oscillators that self-organize toward the onset of synchronization.

The results presented in this work can feed back to both, the experimental exploration of biological systems as well as the design of technological applications. For example, knowing the role of cyclic interaction structures for synchronization gives a hint what to look for in experimental data to explain synchronized behavior, but also what to build into technical devices to achieve it.

The central theme of this work is the analytical treatment of continuous network models. While analytical approaches for discrete networks are well-studied, continuous networks have so far received considerably less attention. Below, we want to contrast both classes of networks, discuss the characteristic difficulties associated with the analysis of continuous networks, and highlight possible approaches.

The dynamics of both, discrete as well as continuous networks, are usually analyzed with the tools of dynamical systems theory. However, the respective dynamical systems are qualitatively different. To see this, consider that in a continuous network, the dynamics of the link and node states can often be described by deterministic differential equations. Thus, the microscopic dynamics of a continuous network directly constitute a deterministic dynamical system. By contrast, describing the dynamics of a discrete network in a dynamical system necessitates a detour: Usually, the dynamics of the discrete link and node states are assumed to be stochastic. Then, the system is described by continuous, coarse-grained variables, whose balance equations are formulated as differential equations.

In summary, a dynamical system description of a discrete network typically pertains to the approximate dynamics of macroscopic variables, while a dynamical system description of a continuous network pertains to the exact dynamics of the microscopic variables. This means that the dynamical system is low-dimensional in case of a discrete network, but high-dimensional in case of a continuous network. Moreover, it implies that the analysis reveals macroscopic properties in the discrete case, but microscopic properties in the continuous case. Thus, studying self-organization in continuous instead of discrete networks imposes two problems: It necessitates to deal with high-dimensional dynamical systems, and to infer the macroscopic properties of interest from the microscopic result of the analysis.

The results presented in this work show that the problems inherent in the analytical treatment of continuous networks can be overcome. Hence, our analysis started in all cases with the consideration of the full, high-dimensional systems. However, we were able to determine in each case model-inherent properties that allowed for the reduction of the dimensionality. Thus, in Chapter 2, the symmetry of the minors with respect to permutations of the index set S disclosed that the cycles of the graph $\mathcal{G}$ can serve as a basis for the graphical notation and calculus. In Section 3, the linear stability analysis revealed that any topological component in the equilibrated network can be characterized by at most four variables. Finally, in Chapter 4, the sequential update rule implied that the process of steering the high-dimensional parameter space can be decomposed in a sequence of one-dimensional processes.

For inferring macroscopic properties from the microscopic-level descriptions we used upscaling procedures. Hence, in Chapter 3, the stationarity and stability conditions first and foremost equated the investments of two links connected by one node, two nodes connected by one link respectively. To scale up the symmetry relations to whole components, we iterated the argument along a sequence of neighbors. In Chapter 2, the analysis followed a similar pattern. Thus, we considered sequences of incrementally growing link chains to derive the topological stability conditions for cyclic subgraphs. Considering multiple subgraphs in turn led to the conjecture of the positive spanning tree criterion. Finally, in Chapter 4, we asked for the prerequisites for an upscaling procedure: We determined under which conditions the value of a microscopic measure O_i allows inferring the value of macroscopic order parameter $\mathcal{O}$.

The challenges associated with the analysis of continuous networks are compensated by certain advantages. Thus, the dynamical system description of a continuous network captures the full topological information and thus allows to study the interplay between structure and dynamics in detail. To describe the structural details of the network on all scales, we complemented the methods from the theory of dynamical systems with methods from graph theory. We interlocked the two toolkits in different ways, usually however by a translation step, in which information about the system is transferred from one framework to the other. To illustrate this point, let us shortly summarize for each project the tie points between the toolkits and the benefits of their interlocking.

In Chapter 2, we interpreted the dynamics related information of the Jacobian $\mathbf{J}$ as adjacency information of a graph $\mathcal{G}$, which allowed to translate the algebraic stability conditions into topological stability conditions. Beside this, the topological interpretation allowed to overcome a technical challenge: For evaluating the necessary algebraic stability conditions the index structure of the terms in a minor was found to be decisive. Describing this structure is intricate in the algebraic framework, but practicable in the graph theoretical framework, which provides suitable concepts and terms such as subgraphs, trees, and cycles.

In Chapter 3, translation between the frameworks allowed to interpret the algebraic conditions for the stationarity and stability of the cooperative investments as topological characteristics of the evolving networks. The tools from dynamical systems side revealed that in equilibrium any topological component can be characterized by at most four variables, which – for stationarity to be observed – additionally have to satisfy one of three possible relations. Arguments from graph theory then revealed that the relations define topologically distinct scenarios, which differ with respect to the average degree in a component as well as to the expected topological arrangements of unidirectional links.

In Chapter 4, finally, the translation step was carried at the very beginning of the considerations, when we mapped a topological structure onto a scalar variable T_i. Therewith, we were able to study the topological self-tuning of the network from the dynamical systems perspective.

In summary, the analytical treatment of continuous network models is in itself more difficult than the analysis of discrete network models. However, it provides results which are exact and account for the detailed topological configuration. The methods derived in this work show ways of overcoming the technical challenges. In the future, it would be desirable to extend them, for example by generalizing the assumptions under which the topological stability analysis introduced in Chapter 2 can be applied. Our hope is that by opening up continuous networks to analytical treatment our methods may contribute to future steps toward a conceptional understanding of adaptive self-organization.

Bibliography

[1] P. W. Anderson (1972) More is different. *Science* 4047 (177): 393–396.

[2] S. A. Kauffman (1993) *The origins of order* (Oxford University Press).

[3] M. Ipsen, and A. S. Mikhailov (2002) Evolutionary reconstruction of networks *Phys. Rev. E* 66: 04610.

[4] B. Derrida, M. R. Evans, V. Hakim, and V. Pasquier (1993) Exact solution of a 1D asymmetric exclusion model using a matrix formulation. *J. Phys. A: Math. Gen.* 26: 1493–1517.

[5] G. Schütz, and E. Domany (1993) Phase transitions in an exactly soluble one-dimensional asymmetric exclusion model. *J. Stat. Phys.* 72: 277–296.

[6] J. T. Chayes, L. Chayes, D. S. Fisher, and T. Spencer (1986) Finite-Size Scaling and Correlation Lengths for Disordered Systems. *Phys. Rev. Lett.* 57: 2999–3002.

[7] A. Donev, S. Torquato, and F. H. Stillinger (2005) Pair correlation function characteristics of nearly jammed disordered and ordered hard-sphere packings. *Phys. Rev. E* 71: 011105.

[8] Y. A. Kuznetsov (1995) *Elements of Applied Bifurcation Theory* (Springer, New York).

[9] A. M. Turing (1952) The Chemical Basis of Morphogenesis. *Phil. Trans. Roy. Soc. London B* 237 (641): 37–72.

[10] A. Gierer, and H. Meinhardt (1972) A theory of biological pattern formation. *Kybernetik* 12: 30–39.

[11] L. G. Morelli, K. Uriu, S. Ares, A. C. Oates (2012) Computational spproaches to developmental patterning. *Science* 336: 187–191.

[12] E. Meron (1992) Pattern formation in excitable media. *Physics Reports* 218: 1-66.

[13] P. Erdös, and A. Rényi (1960) The Evolution of Random Graphs. *Magyar Tud. Akad. Mat. Kutat Int. Kzl.* 5: 17–61.

[14] A. L. Barabási, and R. Albert (1999) Emergence of scaling in random networks. *Science* 286 (5439): 509–512.

[15] R. Albert, H. Jeong, and A. L. Barabási (2000) Error and attack tolerance of complex networks. *Nature* 406: 378–382.

[16] M. E. Newman (2003) The structure and function of complex networks. *SIAM Review* 45: 167–256.

[17] T. Gross, and B. Blasius (2008) Adaptive coevolutionary networks: a review. *J. R. Soc. Interface* 5: 259–271;

[18] T. Gross, and H. Sayama (Eds.) (2009) *Adaptive Networks: Theory, models and applications* (Springer, Heidelberg).

[19] S. Bornholdt, and T. Rohlf (2000) Topological evolution of dynamic networks: global criticality from local dynamics. *Phys. Rev. Lett.* 84(26):6114.

[20] J. Ito, and K. Kaneko (2002) Spontaneous structure formation in a network of chaotic units with variable connection strengths *Phys. Rev. Lett.* 88: 028701.

[21] G. Zschaler, A. Traulsen, and T. Gross (2010) A homoclinic route to asymptotic full cooperation in adaptive networks and its failure. *New J. Phys.* 12: 093015.

[22] P. Holme, and G. Ghoshal (2006) Dynamics of networking agents competing for high centrality and low degree. *Phys. Rev. Lett.* 96, 098701.

[23] F. Vazquez, J. C. González-Avella, V. M. Egu'iluz, and M. San Miguel (2007) Time-scale competition leading to fragmentation and recombination transitions in the coevolution of network and states. *Phys. Rev. E* 76: 46120.

[24] B. Skyrms, and R. Pemantle (2000) A dynamic model of social network formation. *Proc. Natl. Acad. Sci. USA* 97: 9340.

[25] P. Holme, and M. E. J. Newman (2006) Nonequilibrium phase transition in the coevolution of networks and opinions. *Phys. Rev. E* 74: 056108.

[26] M. Lim, D. Braha, S. Wijesinghe, S. Tucker, and Y. Bar-Yam (2007) Preferential detachment in broadcast signaling networks: connectivity and cost trade-off. *Eur. Phys.* 79: 58005–6.

[27] W. Schaper, and D. Scholz (2003) Factors regulating arteriogenesis. *Arterioscler. Thromb. Vasc. Biol.* 23: 1143–1151.

[28] S. Jain, and S. Krishna (2001) A model for the emergence of cooperation, interdependence, and structure in evolving networks. *Proc. Natl Acad. Sci. USA* 98: 543–547.

[29] A. Scirè, I. Tuval, and V. M. Eguíluz (2005) Dynamic modeling of the electric transportation network. *Europhys. Lett.* 71: 318–424.

[30] M. G. Zimmermann, V. M. Eguíluz, and M. San Miguel (2001) Cooperation, adaptation and the emergence of leadership. *Economics with heterogeneous interacting agents* (Springer, Heidelberg), 73–86.

[31] J. Karbowski, and G. B. Ermentrout (2002) Synchrony arising from a balanced synaptic plasticity in a network of heterogeneous neural oscillators. *Phys. Rev. E* 65: 031902.

[32] M. Anghel, Z. Toroczkai, K. E. Bassler, and G. Korniss (2004) Competition in social networks: Emergence of a scale-free leadership structure and collective efficiency. *Phys. Rev. Lett.* 92: 058701.

[33] V. M. Eguíluz, M. G. Zimmermann, C. J. Cela-Conde, and M. San Miguel (2005) Cooperation and the emergence of role differentiation in the dynamics of social networks. *Amer. J. Sociology* 110(4): 977–1008.

[34] S. Goyal, and F. Vega-Redondo (2005) Network formation and social coordination. *Games Econ. Behav.* 50: 178.

[35] A. Grabowski, and R. A. Kosinski (2006) Evolution of a social network: The role of cultural diversity. *Phys. Rev. E* 73: 016135.

[36] J. M. Pacheco, A. Traulsen, and M. A. Nowak (2006) Coevolution of strategy and structure in complex networks with dynamical linking. *Phys. Rev. Lett.* 97: 258103.

[37] C. Zhou, J. Kurths (2006) Dynamical weights and enhanced synchronization in adaptive complex networks. *Phys. Rev. Lett.* 96: 164102.

[38] H. Sayama (2007) Generative network automata: A generalized framework for modeling complex dynamical systems with autonomously varying topologies. *IEEE ALIFE* 07: 214–221.

[39] D. Kimura, and Y. Hayakawa (2008) Coevolutionary networks with homophily and heterophily. *Phys. Rev. E* 78: 016103.

[40] D. H. Zanette, and S. Risau-Gusán (2008) Infection spreading in a population with evolving contacts. *J. Biol. Phys.* 34(1-2): 135.

[41] J. A. Almendral, I. Leyva, I. Sendiña-Nadal, and S. Boccaletti (2010) Interacting oscillators in complex networks: synchronization and the emergence of scale-free topologies. *Internat. J. Bifur. Chaos* 20(3): 753–763.

[42] T. Gross, C. Dommar D'Lima, and B. Blasius (2006) Epidemic dynamics on an adaptive network. *Phys. Rev. Lett.* 96: 208701–4.

[43] A. E. Allahverdyan, and K. G. Petrosyan (2006) Statistical networks emerging from link-node interactions. *Europhys. Lett.* 75: 908–914.

[44] C. Biely, R. Hanel, and S. Thurner (2009) Socio-economical dynamics as a solvable spin system on co-evolving networks. *EPJ B* 67: 285–289.

[45] C. Nardini, B. Kozma, and A. Barrat (2008) Who's talking first? Consensus or lack thereof in coevolving opinion formation models. *Phys. Rev. Lett.* 100: 158701.

[46] F. Vazquez, V. M. Eguluz, and M. San Miguel(2010) Generic absorbing transition in coevolution dynamics.*Phys. Rev. Lett.* 100: 108702.

[47] L. B. Shaw, and I. B. Schwartz (2010) Enhanced vaccine control of epidemics in adaptive networks. *Phys. Rev. E* 81: 046120.

[48] G. Demirel, R. Prizak, P. N. Reddy, T. Gross (2011) Opinion formation and cyclic dominance in adaptive networks. *Preprint* arxiv.org:1011.1124.

[49] I. J. Benczik, S. Z. Benczik, B. Schmittmann, and R. K. P. Zia (2008) Lack of consensus in social systems. *Europhys. Lett.* 82: 48006.

[50] L. B. Shaw, and I. B. Schwartz (2008) Fluctuating epidemics on adaptive networks. *Phys. Rev. E* 77: 0661011.

[51] I. J. Benczik, S. Z. Benczik, B. Schmittmann, and R. K. P. Zia (2009) Opinion dynamics on an adaptive random network. *Phys. Rev. E* 79: 046104.

[52] V. Marceau, P. A. Noël, L. Hébert-Dufresne, A. Allard, and L. J. Dubé (2010)Adaptive networks: coevolution of disease and topology. *Phys. Rev. E* 82: 036116.

[53] E. Volz, S. D. W. Frost, R. Rothenberg, and L. A. Meyers (2010) Epidemiological bridging by injection drug use drives an early HIV epidemic. *Epidemics* 2: 155–164.

[54] G. A. Böhme, and T. Gross (2011) Analytical calculation of fragmentation transitions in adaptive networks. *Phys. Rev. E* 83: 035101.

[55] M. W. Hirsch, and S. Smale (1974) *Differential equations, dynamical systems and linear algebra* (Academic Press, New York).

[56] J. Guckenheimer, and P. Holmes (1983) *Nonlinear oscillations, dynamical systems, and bifurcations of vector fields* (Springer, New York).

[57] E. A. Coddington, and N. Levinson (1955) *Theory of ordinary differential equations* (McGraw-Hill, New York).

[58] D. Zwillinger (1998) *Handbook of differential equations* (Academic Press, London).

[59] Y. Saad (1992) *Numerical methods for large eigenvalue problems* (Manchester University Press, Oxford).

[60] T. Gross, and U. Feudel (2004) Analytical search for bifurcation surfaces in parameter space. *Physica D* 195(3-4): 292–302.

[61] T. B. Benjamin (1978) Bifurcation phenomena in steady flows of a viscous fluid II: Experiments. *Proc. Roy. Soc. A* 359 (1696): 27–43.

[62] D. H. Peregrine, G. Shokert, and A. Symon (1990) The bifurcation of liquid bridges. *J. Fluid Mech.* 212: 25–39.

[63] C. Huepe, G. Zschaler, A. L. Do, and T. Gross (2011) Adaptive network models of swarm dynamics. *New J. Phys.* 13: 073022.

[64] D. Chandler (1987) *Introduction to modern statistical mechanics* (Oxford University Press, Oxford).

[65] H. E. Stanley (1971) *Introduction to phase transitions and critical phenomena* (Oxford University Press, Oxford).

[66] J. J. Binney, N. J. Dowrick, A. J. Fisher, and M. E. J. Newman (1992) *The theory of critical phenomena: An introduction to the renormalization group* (Oxford University Press, Oxford).

[67] J. P. Sethna (2006) *Entropy, order parameters, and complexity* (Oxford University Press, Oxford).

[68] M. Brokate, and J. Sprekels (1996) *Hysteresis and phase transitions* (Springer, New York) 150–163.

[69] E. Ott, and T. M. Antonsen (2008) Low dimensional behavior of large systems of globally coupled oscillators. *Chaos* 18: 037113.

[70] R. Distel (1997) *Graph Theory* (Springer, Heidelberg).

[71] B. Bollobás (1998) *Modern Graph Theory* (Springer, New York).

[72] A. Pikovsky, M. Rosenblum, and J. Kurths (2001) *Synchronization: a universal concept in nonlinear sciences* (Cambridge University Press, Cambridge).

[73] S. Boccaletti (2008) *The synchronized dynamics of complex systems* (Elsevier, Amsterdam).

[74] A. Arenas, A. Diaz-Guilera, J. Kurths, Y. Moreno, and C. Zhou (2008) Synchronization in complex networks. *Phys. Rep.* 469: 93–153.

[75] T. M. Massie, B. Blasius, G. Weithoff, U. Gaedke, and G. F. Fussmann (2010) Cycles, phase synchronization, and entrainment in single-species phytoplankton populations. *Proc. Natl. Acad. Sci. USA* 107: 4236–4241.

[76] L. Glass (2001) Synchronization and rhythmic processes in physiology. *Nature* 410: 277–284.

[77] R. E. Mirollo, and S. H. Strogatz (1990) Synchronization of pulse-coupled biological oscillators. *SIAM J. Appl. Math.* 50 (6): 1645–1662.

[78] W. D. Warner, and C. Leung (1993) OFDM/FM frame synchronization for mobile radio data communication. *IEEE Trans. Veh. Technol.* 42 (3): 302–313.

[79] Y. Kuramoto (1975) *Lecture notes in physics vol. 39* (Springer, New York).

[80] L. M. Pecora, and T. L. Carroll (1998) Master stability functions for synchronized coupled systems. *Phys. Rev. Lett.* 80: 2109.

[81] R. E. Mirollo, and S. H. Strogatz (2005) The spectrum of the locked state for the Kuramoto model of coupled oscillators. *Physica D* 205: 249–266.

[82] C. W. Wu (2007) *Synchronization in complex networks of nonlinear dynamical systems* (World Scientific, Singapore).

[83] Y. Kawamura, H. Nakao, K. Arai, H. Kori, and Y. Kuramoto (2010) Phase synchronization between collective rhythms of globally coupled oscillator groups: Noiseless nonidentical case *Chaos* 20: 0431101–8.

[84] R. Tönjes, and B. Blasius (2009) Perturbation analysis of complete synchronization in networks of phase oscillators. *Phys. Rev. E* 80: 026202.

[85] M. Chavez, D. U. Hwang, A. Amann, H. G. E. Hentschel, and S. Boccaletti (2005) Synchronization is enhanced in weighted complex networks. *Phys. Rev. Lett.* 94: 218701.

[86] T. Nishikawa, and A. E. Motter (2006) Synchronization is optimal in non-diagonizable networks. *Phys. Rev. E* 73: 065106.

[87] I. Lodato, S. Boccaletti, and V. Latora (2007) Synchronization properties of network motifs. *Euro. Phys. Lett.* 78: 28001.

[88] T. Nishikawa, and A. E. Motter (2010) Network synchronization landscape reveals compensatory structures, quantization, and the positive effect of negative interactions. *Proc. Natl. Acad. Sci. USA* 107: 10342–10347.

[89] F. Mori (2010) Necessary condition for frequency synchronization in network structures. *Phys. Rev. Lett.* 104: 108701.

[90] A. E. Motter (2004) Cascade control and defense in complex networks *Phys. Rev. Lett.* 93: 098701.

[91] S. V. Buldyrev, R. Parshani, G. Paul, H. E. Stanley, and S. Havlin (2010) Catastrophic cascade of failures in interdependent networks *Nature* 464: 1025–1028.

[92] A. L. Do, S. Boccaletti, and T. Gross (2012) Graphical notation reveals topological stability criteria for collective dynamics in complex networks. *Phys. Rev. Lett.* 108: 194102.

[93] A. L. Do, S. Boccaletti, J. Epperlein, S. Siegmund and T. Gross (2012) Signature stability analysis for networks of coupled dynamical systems with Hermitian Jacobian *Preprint* arXiv:1207.5699.

[94] J. Epperlein, A. L. Do, T. Gross, and S. Siegmund (2012) Meso-scale obstructions to stability of 1D center manifolds for networks of coupled differential equations with symmetric Jacobian *Preprint* arXiv:1207.3736.

[95] J. A. Acebron, L. L. Bonilla, C. J. Perez Vicente, F. Ritort, and R. Spigler (2005) The Kuramoto model: A simple paradigm for synchronization phenomena. *Rev. Mod. Phys.* 77: 137–185.

[96] X. Liao, and P. Yu (2008) *Absolute Stability of Nonlinear Control Systems* (Springer, Netherlands).

[97] J. M. Beckers (1992) Analytical linear numerical stability conditions for the anisotropic three-dimensional advection-diffusion equation. *SIAM J. Numer. Anal* 29 (3): 701–713.

[98] E. D. Soldatova (2006) Stability conditions for the basic thermodynamic potentials and the substantiation of the phase diagram. *J. Mol. Liq.* 127: 99–101.

[99] J. Cai, X. Wu, and S. Chen (2007) Chaos synchronization criteria and costs of sinusoidally coupled horizontal platform systems. *Math. Probl. Eng.* 2007: 826852.

[100] A. L. Do, L. Rudolf, and T. Gross (2010) Patterns of cooperation: fairness and coordination in networks of interacting agents. *New J. Phys.* 12: 063023.

[101] M. R. Adhikari, and A. Adhikari (2005) *Textbook of linear algebra: Introduction to modern algebra* (Allied Publishers, Mumbai).

[102] G. Kirchhoff (1847) Über die Auflösung der Gleichungen, auf welche man bei der Untersuchung der linearen Verteilung galvanischer Ströme gefhrt wird. *Ann. Phys. Chem* 72: 497–508.

[103] J. Schnakenberg (1976) Network theory of microscopic and macroscopic behavior of master equation systems. *Rev. Mod. Phys.* 48(4): 571–585.

[104] M. Y. Li, and Z. Shuai (2010) Global-stability problem for coupled systems of differential equations on networks. *J. Differ. Equations* 248: 1–20.

[105] G. B. Ermentrout (1992) Stable periodic solutions to discrete and continuum arrays of weakly coupled nonlinear oscillators. *SIAM J. Appl. Math.* 52(6): 1665-1687.

[106] E. M. Izhikevich (1999) Weakly pulse-coupled oscillators, FM interactions, synchronization, and oscillatory associative memory. *IEEE Trans Neural Networks* 10: 508–526.

[107] G. Deffuant (2006) Comparing extremism propagation patterns in continuous opinion models. *JASS* 9 (3): 1460–7425.

[108] G. Eskin, J. Ralston, and E. Trubowitz (1984) On isospectral periodic potentials in $\mathbb{R}^n$ II. *Comm. Pure. Appl. Math* XXXVII: 715–753.

[109] A. M. Shirokov, N. A. Smirnova NA, and Y. F. Smirnov (1998) Parameter symmetry of the interacting boson model. *Phys. Lett. B* 434: 237–244.

[110] I. Sendiña-Nadal, J. M. Buldú, I. Leyva, and S. Boccaletti (2008) Phase locking induces scale-free topologies in networks of coupled oscillators. *PLoS ONE* 3(7): e2644.

[111] J. Almendral,I. Leyva, D. Li,I. Sendiña-Nadal, S. Havlin, and S. Boccaletti (2010) Dynamics of overlapping structures in modular networks. *Phys. Rev. E* 82: 016115.

[112] K. Pyragas (1992) Continuous control of chaos by self-controlling feedback *Phys. Lett. A* 170: 421–428.

[113] R. Toenjes (2007) *Pattern formation through synchronization in systems of nonidentical autonomous oscillators* (Opus, Potsdam).

[114] I. G. Kevrekidis, C. W. Gear, and G. Hummer (2004) Equation-free: The computer-aided analysis of complex multiscale systems. *AIChE J.* 50(7): 1346–1355.

[115] T. Gross, L. Rudolf, S. A. Levin, and U. Dieckmann (2009) Generalized models reveal stabilizing factors in food webs. *Science* 325: 747-750. '

[116] R. Axelrod, and W. D. Hamilton (1981) The evolution of cooperation. *Science* 211: 1390–1396.

[117] M. Doebeli, C. Hauert, and T. Killingback (2004) The evolutionary origin of cooperators and defectors. *Science* 306: 859–862.

[118] R. Gulati, and H. Singh (1998) The architecture of cooperation: managing coordination costs and appropriation concerns in strategic alliances. *ASQ* 43 (4): 781–814.

[119] K. Raustiala (2002) The architecture of international cooperation: Transgovernmental networks and the future of international law. *VA J. Int'l L.* 43 (1): 1–92.

[120] A. Acharya (2011) Asian regional institutions and the possibilities for socializing the behavior of states. *ADB Working Paper* 82: 1–33.

[121] M. Vos, And K. van der Zee (2011) Prosocial behavior in diverse workgroups: How relational identity orientation shapes cooperation and helping. *Group Processes Intergroup Relat.*: 1–17.

[122] J. Ke, J. W. Minett, C. P. Au, and W. S. Y. Wang (2002) Self-organization and selection in the emergence of vocabulary. *Complexity* 7(3): 41–54.

[123] S. Shekar, and D. Oliver (2010) Computational modeling of spatio-temporal social networks: A time-aggregated graph approach. *Proceedings: Spatio-Temporal Constraints on Social Network.*

[124] M. A. Nowak, and K. Sigmund (2004) Evolutionary dynamics of biological games. *Science* 303: 793–799.

[125] M. A. Nowak (2006) Five rules for the evolution of cooperation. *Science* 314: 1560–1563.

[126] R. Axelrod (1984) *The Evolution of Cooperation* (Basic Books, New York).

[127] M. A. Nowak, and R. M. May (1992) Evolutionary games and spatial chaos. *Nature* 92: 826–829.

[128] M. Burtsev, and P. Turchin (2006) Evolution of cooperative strategies from first principles. *Nature* 440: 1041–1044.

[129] C. Hauert, and M. Doebeli (2004) Spatial structure often inhibits the evolution of cooperation in the snowdrift game. *Nature* 428: 643–646.

[130] F. C. Santos, and J. M. Pacheco (2005) Scale-free networks provide a unifying framework for the emergence of cooperation. *Phys. Rev. Lett.* 95: 0981041–4.

[131] H. Ohtsuki, C. Hauert, E. Lieberman, and M. A. Nowak (2006) A simple rule for the evolution of cooperation on graphs and social networks. *Nature* 441: 502–505.

[132] F. C. Santos, M. D. Santos, and J. M. Pacheco (2008) Social diversity promotes the emergence of cooperation in public goods games. *Nature* 454: 213–216.

[133] R. V. Gould (1993) Collective action and network structure. *Am. Soc. Rev.* 58(2): 182–196.

[134] D. Willers (1999) *Network Exchange Theory* (Praeger, Westport, CT).

[135] E. Fehr, and U. Fischbacher (2003) The nature of human altruism. *Nature* 425: 785–791.

[136] G. Palla, A. L. Barabási, and T. Vicsek (2007) Quantifying social group evolution. *Nature* 446: 664–667.

[137] D. Braha, and Y. Bar-Yam (2009) in: *Adaptive Networks* (Heidelberg: Springer), 39–50.

[138] M. W. Macy (1991) Learning to cooperate: Stochastic and tacit collusion in social exchange. *Am. J. Soc.* 97(3): 808–843.

[139] D. Ashlock, M. D. Smucker, E. A. Stanley, and L. Tesfatsion (1996) Preferential partner selection in an evolutionary study of Prisoner's Dilemma. *BioSystems* 37: 99–125.

[140] V. Bala, and S. Goyal (2001) Conformism and diversity under social learning. *J. Econ. Theory* 17: 101–120.

[141] B. Skyrms and R. Pemantle (2000) A dynamic model of social network formation. *Proc. Natl. Acad. Sci. USA* 97: 9340–9346.

[142] M. G. Zimmermann, V. M. Eguíluz, M. San Miguel, and A. Spadaro (2000) Cooperation in an adaptive network. *Adv. Complex Syst.* 3: 283–297.

[143] For a collection of respective publications see http://adaptive-networks.wikidot.com/publications

[144] M. G. Zimmermann, V. M. Eguíluz, and M. San Miguel (2004) Coevolution of dynamical states and interactions in dynamic networks. *Phys. Rev. E* 69: 065102.

[145] V. M. Eguíluz, M. G. Zimmerman, C. J. Cela–Conde, and M. San Miguel (2005) Cooperation and the emergence of role differentiation in the dynamics of social Networks. *Am. J. Soc.* 110(4): 977–1008.

[146] M. G. Zimmermann, and V. M. Eguíluz (2005) Cooperation, social networks, and the emergence of leadership in a prisoner's dilemma with adaptive local interactions. *Phys. Rev. E* 72: 056118.

[147] F. Fu, T. Wu, and L. Wang (2008) Reputation-based partner choice promotes cooperation in social networks. *Phys. Rev. E* 79: 036101.

[148] R. Suzuki, M. Kato, and T. Arita (2008) Cyclic coevolution of cooperative behaviors and network structures. *Phys. Rev. E* 77: 021911.

[149] F. Fu, T. Wu, and L. Wang (2009) Partner switching stabilizes cooperation in coevolutionary prisoner's dilemma. *Phys. Rev. E* 79: 036101.

[150] A. Szolnoki, M. Perc, and Z. Danku (2008) Making new connections towards cooperation in the prisoner's dilemma game. *Euro. Phys. Lett.* 84: 50007.

[151] S. Van Segbroeck, F. C. Santos, T. Lenaerts and J. M. Pacheco (2009) Reacting differently to adverse ties promotes cooperation in social networks. *Phys. Rev. Lett.* 102: 058105.

[152] A. Szolnoki and M. Perc (2009) Resolving social dilemmas on evolving random networks. *Euro. Phys. Lett.* 86: 30007.

[153] J. Poncela, J. Gómez-Gardeñes, L. M. Floría, A. Sánchez, and Y. Moreno (2008) Complex cooperative networks from evolutionary preferential attachment. *PLoS one* 3: e2449.

[154] J. Poncela, J. Gómez-Gardeñes, A. Traulsen, and Y. Moreno (2009) Evolutionary game dynamics in a growing structured population. *New J. Phys.* 11: 083031.

[155] H. Ebel, and S. Bornholdt (2002) Evolutionary games and the emergence of complex networks. *Preprint* arXiv:cond-mat/0211666.

[156] H. Ebel, and S. Bornholdt (2002) Coevolutionary games on networks. *Phys. Rev. E* 66: 056118.

[157] J. M. Pacheco, A. Traulsen, and M. A. Nowak (2006) Active linking in evolutionary games. *J. Theor. Biol.* 243: 437–443.

[158] J. M. Pacheco, A. Traulsen, and M. A. Nowak (2006) Coevolution of strategy and structure in complex networks with dynamical linking. *Phys. Rev. Lett.* 97: 2581031–4.

[159] C. Biely, K. Dragosits, and S. Thurner (2007) The prisoner's dilemma on co-evolving networks under perfect rationality. *Physica D* 228: 40–48.

[160] M. Tomassini, E. Pestelacci, and L. Luthi (2010) Mutual trust and cooperation in the evolutionary hawks-doves game. *Biosystems* 99: 50–59.

[161] A. Szolnoki, M. and Perc (2009) Emergence of multilevel selection in the prisoner's dilemma game on coevolving random networks. *New J. Phys.* 11: 093033.

[162] K. Fehl, D. J. van der Post, and D. Semmann (2011) Co-evolution of behavior and social network structure promotes human cooperation. *Ecol. Lett.* 14: 546–551.

[163] A. L. Do, L. Rudolf, and T. Gross (2012) Coordination, differentiation and fairness in a population of cooperating agents *Games* 3 (1): 30–40.

[164] P. Oliver, G. Marwell, and R. Teixeira (1985) A theory of critical mass. *Am. J. Sociol.* 91: 522–556.

[165] D. D. Heckathorn (1996) The dynamics and dilemmas of collective action. *Am. Soc. Rev.* 61: 250–277.

[166] L. A. Baxter (1984) Trajectories of relationship disengagement. *J. Soc. Pers. Relat.* 1: 29–48.

[167] E. Zeidler, W. Hackbusch, H. R. Schwarz, and B. Hunt (2004) *Oxford User's Guide to Mathematics* (New York: Oxford University Press).

[168] M. D. Koenig, S. Battiston, M. Napoletano, and F. Schweitzer (2008) The efficiency and evolution of R& D networks. *Preprint* CER-ETH Working Paper 08/95.

[169] N. Packard (1988) Adaptation towards the edge of chaos. In: *Dynamic patterns in complex systems* (World Scientific, Singapore), 293.

[170] C. G. Langton (1990) Computation at the edge of chaos: Phase transitions and emergent computation. *Physica D* 42: 12–37.

[171] T. Natschläger, N. Bertschinger, and R. Legenstein (2005) *Advances in neural information processing systems 17* (MIT Press, Cambridge).

[172] J. M. Beggs (2008) The criticality hypothesis: how local cortical networks might optimize information processing. *Phil. Trans. Roy. Soc. A* 366(1864): 329–343.

[173] J. J. Hopfield (1982) Neural networks and physical systems with emergent collective computational abilities. *Proc. Natl. Acad. Sci. USA* 79 (8): 2554.

[174] C. Haldeman, and J. Beggs (2005) Critical branching captures activity in living neural networks and maximizes the number of metastable states. *Phys. Rev. Lett.* 94: 058101.

[175] J. M. Beggs, and D. Plenz (2003) Neuronal avalanches in neocortical circuits. *J. Neurosci.* 23: 11167–11177.

[176] O. Kinouchi, and M. Copelli (2006) Optimal dynamical range of excitable networks at criticality. *Nature Physics* 2: 348–351.

[177] W. L. Shew, H. Yang, T. Petermann, R. Roy, and D. Plenz (2009) Neuronal avalanches imply maximum dynamic range in cortical networks at criticality. *Journal of Neuroscience* 29(49): 15595–15600.

[178] W. Freeman, L. Rogers, M. Holmes, and D. Silbergeld (2000) Spatial spectral analysis of human electrocorticograms including the alpha and gamma bands. *J. Neurosci. Methods* 95: 111–121.

[179] J. M. Beggs, and D. Plenz (2004) Neuronal avalanches are diverse and precise activity patterns that are stable for many hours in cortical slice cultures. *J. Neurosci.* 24: 5216–5229.

[180] E. Gireesh, and D. Plenz (2008) Neuronal avalanches organize as nested theta and beta/gamma-oscillations during development of cortical layer 2/3. *Proc. Natl. Acad. Sci. USA* 105: 7576–7581.

[181] T. Petermann, T. C. Thiagarajana, M. A. Lebedevb, M. A. L. Nicolelisb, D. R. Chialvoc, and D. Plenz (2009) Spontaneous cortical activity in awake monkeys composed of neuronal avalanches. *Proc. Natl. Acad. Sci. USA* 106: 15921–15926.

[182] P. Bak, C. Tang, and K. Wiesenfeld (1987) Self-organized criticality: an explanation of $1/f$ noise. *Phys. Rev. Lett.* 59: 381.

[183] K. Christensen, R. Donangelo, B. Koiller, and K. Sneppen (1998) Evolution of random networks. *Phys. Rev. Lett.* 81: 2380.

[184] M. Paczuski, K. E. Bassler, and A. Corral (2000) Self-organized networks of competing boolean agents. *Phys. Rev. Lett.* 84: 3185–3188.

[185] S. Bornholdt, and T. Röhl (2003) Self-organized critical neural networks. *Phys. Rev. E* 67: 066118.

[186] A. Levina, J. M. Herrmann, and T. Geisel (2007) Dynamical synapses causing self-organized criticality in neural networks. *Nature Physics* 3: 857-860.

[187] T. Rohlf (2008) Self-organization of heterogeneous topology and symmetry breaking in networks with adaptive thresholds and rewiring. *Euro. Phys. Lett.* 84(1): 10004.

[188] Z. Burda, A. Krzywicki, and O. C. Martin (2008) Adaptive networks of trading agents. *Phys. Rev. E* 78(4): 046106.

[189] C. Meisel (2009) Self-organized criticality in adaptive neural networks. *Diploma thesis* http://www.biond.org/node/91

[190] C. Meisel and T. Gross (2009) Self-organized criticality in a realistic model of adaptive neural networks. *Phys. Rev. E* 80: 061917.

[191] B. D. MacArthur, R. J. Snchez-Garca, and A. Ma'ayan (2010) Microdynamics and criticality of adaptive regulatory networks. *Phys. Rev. Lett.* 104: 168701.

[192] S. S. Manna (1991) Two-state model of self-organized criticality. *J. Phys. A* 24: L363.

[193] P. Bak (1996) *How nature works: The science of self-organized criticality* (Copernicus, New York).

[194] A. Vespignani, and S. Zapperi (1998) How self-organized criticality works: a unified mean-field theory. *Phys. Rev. E* 57(6): 6345.

[195] H. Jensen (1998) *Self-organized criticality* (Cambridge University Press, New York).

[196] D. L. Turcotte (1999) Self-organized criticality. *Rep. Prog. Phys.* 62: 1377 –1429.

[197] R. Dickman, M. A. Muoz, A. Vespignani, and S. Zapperi (2000) Paths to self-organized criticality. *Braz. J. Phys.* 30 (1): 27–40.

[198] R. W. Williams, and K. Herrup (1988) The control of neuron number. *Annu. Rev. Neurosci.* 11: 423–53.

[199] J. A. Bonachela, and M. A. Muñoz (2010) Self-organization without conservation: true or just apparent scale-invariance? *J. Stat. Mech.* 02: P02015.

[200] B. Ewing, and P. Green (2000) Analysis of expressed sequence tags indicates 35000 human genes. *Nature Genetics* 25, 232–234.

[201] G. Innocenti, and D. Price (2005) Exuberance in the development of cortical networks. *Nature Rev. Neurosci.* 6: 955–965.

[202] D. Price, H. Kennedy, C. Dehay, L. Zhou, M. Mercier,.Y. Jossin, A. M. Goffinet, F. Tissir, D. Blakey, and Z. Molnr (2006) The development of cortical connections. *Eur. J. Neurosci.* 23: 910–920.

[203] B. Walmsley, A. Berntson, R. Leao, and R. Fyffe (2006) Activity-dependent regulation of synaptic strength and neuronal excitability in central auditory pathways. *J. Physiol.* 572: 313–321.

[204] G. Turrigiano, K. Leslie, N. Desai, L. Rutherford, and S. Nelson (1998) Activity dependent scaling of quantal amplitude in neocortical neurons. *Nature* 391: 892–895.

[205] H. Markram, J. Lbke, M. Frotscher, and B. Sakmann (1997) Regulation of synaptic efficacy by coincidence of postsynaptic APs and EPSPs. *Science* 275 (5297): 213–215.

[206] K. E. Kürten (1988) Critical phenomena in model neural networks. *Phys. Lett. A* 129: 157–166.

[207] J. B. Rundle, K. F. Tiampo, W. Klein, and J. S. Sá Martins (2002) Self-organization in leaky threshold systems: The influence of near-mean field dynamics and its implications for earthquakes, neurobiology, and forecasting. *Proc. Natl. Acad. Sci. USA* 99: 2514–2521.

[208] K. F. Tiampo, J. B. Rundle, W. Klein, J. S. Sá Martins, and C. D. Ferguson (2003) Ergodic dynamics in a natural threshold system. *Phys. Rev. Lett.* 91: 238501.

[209] J. G. T. Zañudo, M. Aldana, and G. Martínez–Mekler (2011) Boolean threshold networks: Virtues and limitations for biological modeling. *Information Processing and Biological Systems* (Springer, Berlin): 113–151.

[210] S. Assenza, J. Gómez-Gardeñes, V. Latora, and S. Boccaletti (2011) Emergence of structural patterns out of synchronization in networks with competitive interactions. *Preprint*

[211] P. Erdös, and A. Rényi (1959) On random graphs I. *Publicationes Mathematicae* 6: 290–297.

[212] R. Barrio (2005) Performance of the Taylor series method for ODEs/DAEs. *Appl. Math. Comput.* 163(2): 525–545.

Index

adaptive network, 22, 28, 46, 56
adjacency matrix, 6, 16

bifurcation, 3

cellular automaton, 53, 55
codimension, 3, 61, 73
complex network theory, 5
component, 6, 16, 40, 43
continuous network, 6, 76
continuous phase transition, 4, 55
cooperation, 27
coordination, 27, 31, 50
criticality, 4, 53, 67
cycle, 6, 12, 44

degree, 7, 39, 48
differentiation, 27, 46, 50
directed network, 6, 28
discrete network, 28, 57, 76
dynamical system, 1, 39
dynamical systems theory, 2

eigenvalue, 3, 11
emergence, vii

graph theory, 5, 39
graphical notation, 11, 13

homeostatic plasticity, 57, 58
homogeneity, 64, 66, 69

inflection point, 34, 38
integrate-and-fire neuron, 57

isospectrality, 22

Jacobi's signature criterion, 3, 10, 21, 36
Jacobian matrix, 2, 10, 34

Kirchhoff's theorem, 14
Kuramoto model, 10, 20, 68

locality, 7, 59
long-term behavior, 2, 57

mesoscale, 10, 24
minor, 11

non-linear, 2, 29

order parameter, 5, 55, 56, 62, 69
ordinary differential equation, 2, 30, 66

parameter space, 3, 56, 73
path, 6
phase locked, 10, 21, 68
phase oscillator, 10, 24, 68
phase space, 1
phase transition, 4, 41, 53, 60
phases, 4, 53, 66
power-law, 4, 54, 55
prisoner's dilemma, 28

random graph, 40, 61, 70
robustness, 57, 72

scale invariance, 54

About the authors

Ly Do has been trained in mathematics, physics and voice. She started her career as a soprano but then decided upon a reputable life as theoretical physicist. During her PhD, Thilo Gross kindled her enthusiasm for selforganization problems, trained her to see adaptive networks everywhere, and encouraged her to tackle them with math. Recently, Ly has become head of the visitors program of the Max Planck Institute for the Physics of Complex Systems, a position that will allow her to further probe the blank spots on the map of adaptive self-organization.

Thilo Gross studied Physics in Oldenburg and Portsmouth. After completing his PhD in 2004 he worked in Potsdam and Princeton before joining the Max-Planck-Institute for the Physics of Complex Systems as group leader in 2007, where he was fortunate to supervise Ly Do as his first PhD student. In 2011 he joined the University of Bristol, where he presently holds a Readership in the Merchant Venturers School of Engineering. His work focuses on the analysis of complex systems with tools from nonlinear dynamics and network science. He is interested in almost everything.